C000269066

GOODBYE BERLIN

GOODBYE BERLIN

THE BIOGRAPHY OF GERALD WIENER

MARGARET M. DUNLOP

This edition first published in 2017 by
Birlinn Limited
West Newington House
10 Newington Road
Edinburgh
EH9 1QS

First published in hardback in 2016

www.birlinn.co.uk

Copyright © Margaret M. Dunlop 2016

Foreword copyright © Grahame Bulfield 2016

The moral right of Margaret M. Dunlop to be identified as the author of this work has been asserted by her in accordance with the Copyright, Designs and Patents Act 1988.

All rights reserved.
No part of this publication may be reproduced, stored or transmitted in any form without the express written permission of the publisher.

ISBN 978 1 78027 541 3

British Library Cataloguing-in-Publication Data
A catalogue record for this book is available from the British Library

Typeset by Initial Typesetting Services, Edinburgh
Printed and bound by Gutenberg Press, Malta

Contents

List of Illustrations

Foreword

I have known Gerald Wiener as a geneticist for around fifty years. His research on the genetics of copper absorption and its role in swayback disease in sheep was groundbreaking, opening up a new area of physiological genetics of farm animals. In the research group he assembled around himself in Edinburgh were several young scientists who later became internationally famous researchers themselves, including Sir Ian Wilmut, leader of the team that cloned Dolly the sheep.

What I had not realised was the dramatic back-story to Gerald's life – escaping from Hitler's Germany on one of the last Kindertransports; being almost alone at twelve in England; looked after by a variety of people, some of whom showed him exceptional kindness; his discovery of a life-long interest in farm animals; and his determination to get to university. This he did, graduating from Edinburgh in 1947, and he remained there, latterly at the Roslin Institute, for the rest of his career. Later in life Gerald managed to find many members of the family he had left behind in Germany, including two brothers, in the USA, bringing a partially happy resolution to a dreadful twentieth-century story.

When Gerald retired as an active scientist he became a consultant for the Food and Agriculture Organization of the United Nations (FAO), advising on cattle and sheep breeding in some challenging areas of the world, including Yemen, Ethiopia and North Korea. He also had long-term involvement in animal breeding programmes in India and China, where for many years he advised on yak breeding, especially in Tibet and South-west China. Gerald is the author of two books, one on tropical animal breeding and the other on yaks, the latter making him a recognised world expert on yak breeding.

For a research scientist, Gerald has had a difficult, challenging and finally rewarding life. This biography of a remarkable journey has been written in a flowing and intriguing narrative by the author Margaret M. Dunlop, who has accompanied Gerald on the latter part of it. It is a story of triumph over terrible adversity to international success.

Professor Grahame Bulfield, CBE, DSc, FRSE:
Formerly Director and Chief Executive of Roslin Institute and
Vice-Principal and Head of Science and Engineering at the
University of Edinburgh

Author's Preface

This book is the story of a remarkable life of perseverance and success that began in the most inauspicious of circumstances in Nazi Germany. It is the biography of Professor Gerald Wiener BSc, PhD, DSc, FRSE, CBiol, FRSB, a scientist of some note but also a man of many other interests including, for most of his life, his involvement in community and church affairs.

The material for this biography derives from the recollections of the man himself, obtained through extensive discussion, and also recollections from members of his family, including, importantly, his mother, prior to her death. In addition, colleagues and friends of Gerald Wiener contributed and I am grateful to them all. Most of the information came from these discussions but there were also written and voice-recorded notes and reports from Gerald, and letters from him and from members of his family.

As Gerald Wiener's wife in his later years I have had privileged access to all of this and was also able to add my own recollections. I persuaded him that his story had to be told, as it is both unusual and uplifting. After his initial reluctance to be thrown into the public eye in this way he gave his support and encouragement to this venture. Throughout the book I have refrained from writing in the first person about myself, so the 'Margaret' who appears in the opening chapter and in the latter half of the book is portrayed essentially as a bystander, as she would have been had the biography been written by another. I have used my pen name rather than my married name as I have two other books to my credit.

By way of acknowledgement I also wish to thank my two editors for helpful comments and suggestions. Any errors, however, are mine alone. I also want to mention that the staff of my publishers could not have been more helpful, most especially Andrew Simmons, editorial manager, for his patience and unfailing courtesy.

Margaret M. Dunlop

Chapter 1

Hello, Dolly

The sun was shining that day. It shone down on the castle, the great fortress of Edinburgh, perched high above the town. It warmed the milling crowds on Princes Street, the tourists and happy shoppers. It was a Friday in July 1997 and the mood of the city was light. In a café opposite the McEwan Hall of the University of Edinburgh the sun shone down upon a table where, blinking in the light, a clutch of distinguished-looking older citizens, grizzled and serious, were seated. They were part of a group who had graduated from that same university fifty years before, to the day.

The great city of Edinburgh is famed for its history in the world of education, and of modern ideas in politics and economics. It was referred to as a 'hotbed of genius' by the novelist Tobias Smollett. That was in the eighteenth century and some would say it hasn't changed. The famous people of Edinburgh's past make up an amazing list, including naturalist Charles Darwin, philosopher David Hume, surgeon Joseph Lister, inventor Alexander Graham Bell, writers Arthur Conan Doyle, R. L. Stevenson, J. M. Barrie and Sir Walter Scott.

On this day, from the doors of the McEwan Hall was streaming a bevy of young men and women, black-robed and smiling as they clutched their degree certificates. They stood in the sunlight, chattering with happiness under the gaze of friends and doting parents.

The contented older group sipped their coffee, their attention slowly turning to the scene on the other side of the road. Through the passing cars and pedestrians they could just catch glimpses of the joyful scenes taking place. They looked at each other and smiled. They had gathered that day in Edinburgh to celebrate. It was a reunion organised by Dr Gerald Wiener along with friends Ken Runcie and Barbara Findlay from the class of '47. These seemingly ancient graduates, all former students of agriculture, were revisiting their former haunts in the city. One of the seated men called out to a group of young graduates, 'Hey, we're celebrating our graduation ceremony of fifty years ago!' The students slowed, smiling in disbelief. The old man continued, 'I'm from Iceland, you know, and this guy here came all the way from Canada to be here today.' The young folk laughed and hurried on. How could these friendly old folk be compared to themselves? Today was theirs. Everyone knew that.

Out of the eighteen students who had graduated in 1947, only eleven had come to the reunion. A few had brought their wives, including Gerald, who had brought Margaret. She was just as excited as the others about the meetings of that day. Later, the celebrating old graduates were to be in for a big surprise, though they didn't yet know it.

They left the café and headed to the university's Old College quadrangle, that beautiful, classical-style building designed by Robert Adam. It was a start to bringing back more memories. There they climbed into a minibus and began their journey. Turning along Princes Street, the Scott Monument came into view, rising high and stark to the left of the bus. 'I once climbed the staircase to the top,' Ken Runcie boasted proudly, 'and got a certificate for my pains. But I tell you – I couldn't do it now.' Most of the party thought they would not even have attempted the 287-stair climb. They passed alongside Princes Street Gardens, glancing up at the iconic castle, which dominated the city. Next they passed the Usher

Hall, where some of them had gone to concerts. On the other side of the road were impressive modern office blocks that the group had not seen before. These were the new symbols of Edinburgh as a financial centre. The bus turned left along the Meadows – perhaps to remind them of a walk home with a girlfriend from an 'Agri Soc' dance at the Students' Union – and then past King's Buildings, the science campus of the university where so much of their time had been spent as students. So engrossed were these men and women in the journey and the conversations among themselves that they sat up with a jolt when the bus pulled up outside the Dryden Field Station of the Animal Breeding Research Organisation in Roslin, a few miles west of Edinburgh. Gerald Wiener was getting excited, but he restrained himself from reminding the party that this had been his department. Here he had spent hours of his life on engrossing research.

The party was shown into a shed to be faced by a sheep, looking straight at them through the bars of its straw-lined pen. It was as though the sheep was saying hello to a new set of friends. She had become quite accustomed to visitors. All members of the party had heard about this sheep. How could they not? 'Dolly' was world-famous as the first mammal in the world to be cloned from an adult cell. And what possible response could there be to a sheep so clearly looking at them? The Canadian glanced wryly at the rest of the group and smiled as he leaned against the pen, saying, 'Well, hello, Dolly!'

Ian Wilmut (later knighted as leader of the team of this momentous achievement) had been called away at the last moment and was unable to greet the party, but one of his collaborators, Bill Ritchie, was there to welcome the group and talk about the cloning project.

Gerald, as a former colleague of Ian, had been able to arrange this privileged access to Dolly, then just a year old. Visitors were tightly restricted but Dolly was a pampered animal and probably more used to human company than to fellow sheep. Gerald stepped in front of the group to provide an introduction to the occasion.

He had invited his friends on this, the fiftieth anniversary of their graduation, to cast their minds back to 1945 and 1946 when they had learned their animal genetics. The genetics then learned from Hugh Donald, though still relevant today, had changed out of all recognition, particularly the way in which the genotypes of animals can be manipulated and reproduced. This was now a field undreamed of fifty years ago and he wondered with what mixture of enthusiasm and scepticism Donald would have talked about it were he still alive. He then went on to introduce Bill Ritchie, a man at the leading edge of this new animal genetics and in fact the one without whom there might have been no 'Dolly'. Bill was the man who had performed the intricate microscopic manipulations leading to the birth of the cloned sheep. Bill started by answering the question on everyone's mind: 'Why the name Dolly?'

'The cell from which this sheep was cloned,' he said, 'came from the preserved mammary gland of a Finn-Dorset ewe. Dolly Parton, the famous country and western singer, flashed into Ian Wilmut's mind as someone well endowed in her upper region – it was just an irreverent spur of the moment thought, but the name Dolly stuck'.

On a more serious note, Bill told the interested company that cloning from an adult cell was far more difficult than doing so from embryonic cells, which had been done before. Dolly was the only lamb born from more than 270 attempts where the scientific team had had success. But the purposes for which the cloning was intended required this procedure. It was done in support of other research to produce sheep milk containing human genes that would produce proteins for use in medicines. This, for example, could be used to treat haemophilia. The process of inserting new genes into sheep for their expression in milk was itself extremely difficult and, once achieved, many such sheep would be needed to provide enough milk to be commercially useful. Cloning would therefore be a way of replicating an animal that had the new proteins in sufficient quantity in its milk. There was never a thought that cloning might be extended to humans, as was quickly speculated in more lurid publications – and of course any such extension of cloning for humans was just as quickly banned by governments.

The retired scientists were grateful for being among the privileged few to have seen the famous animal in the flesh. Dolly, still peering through the iron bars of her pen, looked as if she were wishing to say goodbye to her guests, they thought.

Lunch was provided in the splendid Bush House, which had been the mansion house of the large estate that the university had purchased years earlier. Now the house was a centre for final year and post-graduate students, with a good canteen, study rooms and student accommodation. Murray Black, the overall farm manager, still living at Boghall, the original college farm that all the former students had known well, provided the welcome and guided the eager and well-fed party over the estate and its work.

There was one last visit in store that day on the return to Edinburgh. At the grand buildings of the School of Agriculture on the King's Buildings site of the university, Peter Wilson, Professor of Agriculture and Principal of the College of Agriculture, was there to welcome the former students. They thought back to the old location in George Square where they had studied and were amazed by these seemingly vast premises. But Peter Wilson, sensing the question, explained that there had also been a large expansion in the advisory and investigative functions to the industry for which the college had responsibility and which were also housed in these buildings.

Gerald and Margaret returned with two of the party, Ólafur Stefánsson and his wife, Thorum, to their home in Biggar where these two Icelanders were staying as guests. Ólafur had been one of Gerald's best friends during their student days. He was also one of a distinguished line of Icelandic men and women who had studied agriculture in Edinburgh, going back as far as 1887. Ólafur himself had risen to top roles in Icelandic's agriculture sector. He became acting Director of Agriculture in Iceland. The two friends had kept in touch by letter for many years after graduation. Apart from Ken Runcie and, briefly, Barbara Findlay when organising the meeting, Gerald had not met any of the party since their graduation fifty years earlier, nor were they to meet again.

After graduation Ken Runcie had become a university lecturer before rising to a top position in the School of Agriculture. Most of the others had gone back to farming lives or moved to other agricultural professions. The Canadian, Andy Lyle, had returned to his own country. Everyone thought it had been a rewarding trip, and an emotional occasion in reliving old memories.

That evening, over a glass or two, Ólafur and Gerald and their wives got to talking about the past. Ólafur said, 'Did you ever think your life would work out so well, Gerald, when you were a boy, working as a labourer on a farm and watching every penny that you earned? What a climb you have made! Deputy Director of the Research Organisation, and could have been director, so I'm told!'

Gerald smiled, saying that that had been a situation that arose long ago in his career. He was not a great drinker but the day's events had got to him and he sipped at his whisky, saying, 'No, I never did, but it's all in the past now. It's been a fascinating life. I'm sure that is true of all us visiting Dolly today. It's been a wonderful career that I've had, and more than that I now have friends and colleagues all over the world, thanks to the University of Edinburgh, and the great teaching, the first-class treatment all we students had. What a privileged lot of people we have been.'

'A toast!' said Margaret. 'To Edinburgh and Scotland!'

'To England and Oxford!' said Gerald.

'To Great Britain!' said Ólafur.

Chapter 2

Berlin Childhood

He was called Horst then, in 1930s Germany. He was in Berlin with his mother. Happily out together for an afternoon's shopping, they were in the food hall of the KaDeWe, the large department store in the west end of the city. It was December 1936, and Sunday would be his grandfather's birthday. His mother, Luise Wiener, had taken her little boy there for a special purchase. The whole of the food hall at that time was a feast for the eye. There were counters laden with fruits and vegetables, cheeses and sausages from all over the world. There were coffees, teas, spices and herbs, a butcher's counter for fresh meat and, best of all for ten-year-old Horst, the large tanks with fish of all sorts swimming around as in an aquarium. He pictured how he might come into sight to these enormous fish as he gazed at them in fascination. One of the creatures flipped its tail as a gloved finger appeared in front of Horst's face. It was the finger of Liesl (the affectionate, familiar name for Luise).

She had picked out a fine, big carp, her father's choice as a treat on his birthday. Asking the fishmonger to stun the creature, she explained that it wouldn't be needed for two days, until the birthday party. He gave the fish a slanted blow to the head and advised her to keep it in cold water in the bath.

Luise took the boy's hand, having placed the fish, wrapped in a damp cloth, in her shopping bag. Together they started out towards

Bamberger Strasse. On meeting a neighbour, Helga Plowdiski, the lady from the stationery shop beneath their flat, they stopped and talked of the forthcoming birthday party. They laughed together as Helga remarked that Herr Trost, Luise's father, was handsome and that her own mother had a fancy for him. They went on to talk of local affairs, Helga asking about Luise's family. Horst was getting impatient. He was worried about the poor fish. He longed to get it back to the apartment so that they could put it in the bath, and he tugged at his mother's skirt, pleading quietly, 'The fish! The fish!' Soon Luise relented. The two young women said goodbye.

Luise called back to Helga to suggest that she come up on Tuesday if her mother gave her time off from the shop. They could have some coffee, and Luise wanted to show her some old photographs of Horst when he was younger and of Paul, her ex-husband.

Then Luise and Horst walked on. For twenty minutes they walked past building after building, then up stairs, until the golden moment for Horst arrived and the great heavy carp was released into the cold water of the bath.

On Sunday, Luise's sister, Erna, and her husband, Erich, arrived with Horst's cousin, Marion – a good friend and playmate, almost a sister. Both were only children. The fish had gone from the bath, and the children played in one of the bedrooms while the adults talked.

It was a pleasant, spacious flat in the western part of Berlin. The living room opened out with French doors on to a balcony. The overhang from the balcony above shut out a lot of the light and gave the room a somewhat sombre air. The large mahogany dining table with its six chairs stood in the centre, dominating the room. There was a large armchair usually occupied by Grandfather, where he read his newspaper or listened to the radio. There was also a settee – a dark, heavy piece of furniture, but comfortable. In one corner of the room was a large stove, stretching from the floor to just a little short of the ceiling. It was tiled green on the outside and was the principal source of heat for the flat. The warming oven in the upper half of the stove was rarely used. Leading off the living room was Grandfather's bedroom and another door led onto the passage with the kitchen opposite and two other bedrooms, one for Luise and one for Horst.

The table had been set for the birthday party, the living room cleaned and polished in preparation for the celebrations. The tempting aroma of cooking was everywhere in the apartment. Anticipation of the Sunday meal, special today, was in everyone's mind, but first the adults had to have their usual discussion of family affairs. These conversations started in a friendly fashion but went up a notch in tone when matters of money were raised. And then there was politics, and the subdued talk of discrimination against Jews and of the frightening stories of concentration camps.

On the evening of the party for Horst's grandfather, Hermann Trost, one could still hear the sound of raised voices and angry tones. But the children had become used to the voice of irritation from Hermann. Regularly he berated his son-in-law Erich, and regularly he would then turn to his older daughter, Erna, and scold her for the way they spent their money unwisely, such as the expensive-looking outfit Marion was wearing that day.

For a few minutes, the old man's mind went back to the hard days of the past. Hermann had known tough times in his youth. At the age of fourteen he had walked all the way from Poland, a journey of many days and many nights, seeking work in Germany. After some years of saving his wages he and his wife had set up a drapery and fabric shop in the town then known as Liegnitz in Silesia. (This was before Liegnitz was handed back to Poland after the Second World War and renamed Legnica). The shop became a good-going business. They sold heavy checked cotton material, stuff that would withstand the work in the fields, to farmers' wives making shirts for their husbands, and finer material for dresses for the wives themselves. Hermann was a good business man. He and Sarah, his wife, scrimped and saved and gradually became successful. Now the old man could not lose his habit of watching his money.

Erna was two years older than Luise. She had married Erich, who was a commercial traveller and therefore had a car – a rare thing in those days. He had fought on the German side in the First World War and was decorated with the Iron Cross for bravery. However, in Hermann's eyes they wasted their money on fripperies and they had to withstand a lecture every Sunday when they called to see the old man.

The cousins played on, consciously ignoring the angry voices. Always about money. Horst did not remember his father. He remembered nothing of the little town of Küstrin where he was born. One day he would learn the story of how Paul Wiener had met his mother at a dance in the local village. Of how Paul, debonair and wild, had kissed her at the door and she had said, 'Now you have kissed me, you will have to marry me!' Well that's how the story goes. The year was 1924. Luise was very beautiful, slim and romantic-looking in those days as seen from her photographs. Her father was prosperous.

With money from Luise's father, Paul was set up in a men's and boys' outfitters shop in the town of Küstrin in Silesia (now called Kostrzyn and ceded to Poland after the Second World War). The young couple's shop was several miles away from Luise's parents' house in Liegnitz. The Wieners came from Breslau (Wrocław) originally. Luise's father's idea was to give the young couple a start in making a living. But things did not work out. Paul was a natural musician. He played both the piano and the piano accordion very well, and had recently started his own little dance band. He was also very artistic and, in Luise's eyes, a bit reckless. He liked stunts, some of which were, of course, to advertise their business. One day he brought back a lion cub from a travelling circus to put in the shop window to attract passers-by. Luise did not approve, especially with their baby in the rooms just behind the shop. But more unacceptable than stunts and a preference for his musician life over the shop was the gossip about Paul. Soon after the birth of his little son it became common knowledge that Paul had not one but two girlfriends in the local area. A proper Romeo. Poor Luise was devastated. Soon she had had enough of his nights out and his unfaithfulness. Always out with the orchestra, as he called it. Often seen with glamorous women. His heart and mind were always elsewhere. Horst was only two years old when Luise took him and returned to live with her parents in Liegnitz and help in their shop. She had tried, but in the end, following a period of separation, divorce had to be accepted two years later.

When Luise's mother died after a short illness, Hermann decided to sell up his business, to move into retirement in Berlin and to take Luise and his grandson with him. On the night of Grandfather's birthday party, Horst and his mother had been living with him for three years. This was a sadder, quieter way of life for the young Luise, still only in her early thirties. Here in the big city, she did not have to work in the draper's shop, and she did not have any money worries, as her father was very good to her. She had a woman to do the cleaning and laundry for her, but still she missed having a partner, a husband who would care for her, and love her. It was sad for Luise to feel alone, and sad that her mother, whom she adored, had passed away. But she had her father, and her beautiful son. The boy could be educated in Berlin while she could take care of her ageing father. And Horst was a great comfort and joy to her. Like his handsome father, he had a bright and sunny nature, a dazzling smile, and eyes full of mischief.

CHAPTER 3

Early Schooldays

It was the first day of school for Horst. Along with dozens of boys and girls, he trooped off on the five-minute walk from where he lived, accompanied by his mother. Outside the school was a throng of the young pupils and their doting parents, excited by this special day. As was the custom, each child had been given a large paper cone full of sweets. Each child was cuter than the next; with their brand new clothes and innocent faces they trooped into their classroom carrying their colourful cones. And at this school Horst stayed for two years. But, sadly, this time came to an end when one day Luise was called into the Headmaster's room.

'Frau Wiener, we are sorry to tell you that we have been instructed by the authorities that we can no longer have Jewish children in our school.' The man was clearly embarrassed at having to convey this message, but continued in regretful tones, 'I will have to ask you not to bring Horst back to this school after today.'

Horst was enrolled in a private Jewish school, the 'Lessler Schule', where he continued quite happily, being a bright child and quick to learn. Here he met his first and best friend, Hardy Seidel. They were friends for almost eighty years, close as brothers right up until the time of Hardy's death in London at the age of eighty-six in 2010. Horst and his cousin, Marion, were like brother and sister. Each Sunday, they met and played together in those carefree years of their childhood, ignoring the rising voices of the adults as the atmosphere in Berlin gradually worsened. They were both sent to a holiday camp one summer, the idea being for them to get a break from their worried parents and to be outdoors most of the day. They were weighed weekly to see if they had gained weight and the one that gained most got a prize. As Horst was always rushing around he was the only child who gained no weight at all.

At home he had his good friend, Hardy. Each day after school they would take their little model cars and race them around the low wall surrounding the fountain at the end of the street. Sometimes they would call at the little newsagent's shop for pencils and comics. There they chatted with the old lady and sometimes with her daughter, Helga, who were neighbours and friends of Luise.

One day, when the two of them were sitting on the wall of the fountain, silent for a change, Hardy in stumbling tones informed Horst that he thought he would be going away from Berlin quite soon – probably to America. Horst's heart sank. He listened while Hardy explained that his mother had been so upset and angry when she couldn't sit down in the park because of the notices that had been put up by the benches saying 'No Jews Allowed'. It was then that the Seidels decided that enough was enough.

The two boys were silent for several minutes, slowly dealing with their thoughts. Horst's mind was on his grandfather's words to his mother: 'Be quiet, Liesl, you'll see it will all blow over. I've seen it all before.'

Hardy said that his cousin had been attacked by a gang of thugs in uniforms on the corner as he had walked past.

Slowly, it was dawning on Horst that he was about to lose his best friend. Life had been so good, so easy, so sunny. But now the clouds were moving in.

The Seidels did not go to America after all, but took up an offer to go to London instead. Horst's grandfather, it seemed, helped them a little financially. Hardy's father had been in a panic about the deteriorating situation around them. He was more politically aware than his neighbours, and the danger signals of each passing day had scared him and they left abruptly. Luise and her father heard from the Seidels in due course. They were doing all right in London, and they urged Hermann Trost to think about moving. But the old man turned a deaf ear. The year was 1937.

Helga Plowdiski had stories for Luise of her mother's worry about their little stationer's shop. Business was very bad. The burly boys of the Hitler Youth had taken to standing outside their premises, stopping people from going in, saying, 'These are Jews, you know. Don't give them your money. They have brought all the trouble to our country.' Another day she came to the apartment on the second floor with tears in her eyes. Her uncle who was a doctor had been told to close down his practice and move somewhere else. No longer were Jewish people allowed to be doctors, or to sell medicines. The poor young woman was close to a breakdown. Luise was scared. Each day she tried to persuade her father to leave Berlin. But it was no use, he would not listen. She began to make her own plans.

As is well known, things got worse and worse for Jews in Germany. There was at this time a dreadful happening, a day never to be forgotten. The residents of their little friendly street were stunned. It was November 1938, the awful Night of Broken Glass, 'Kristallnacht'. The Nazis had broken all the windows on any shops in the street that were owned by Jews, and painted 'Jude' on the doors. This included the little stationer's shop owned by Luise's friends. It was a shocked Luise who saw the destruction next morning as she looked out from the balcony of their apartment. Suddenly Horst, now twelve years old, became fully aware of the danger of being Jewish in Berlin at that time. Although they did not attend synagogue, had no kosher rules in the house, and never discussed religion, they were to be outcasts in their own country. It was about then that Luise wrote to the Seidels in London to ask for help. She had heard of the efforts of the Save the Children organisation. She asked if they could get in touch with this organisation to help Horst to escape from the dreadful situation building up in Berlin.

Luise had to carry on with her domestic duties, looking after her son and her father. She had to hide her anxiety from both of them. She had to keep in check the sadness that would come when she had to say goodbye to both of them. At last the papers came through, and Horst was to be allowed to go unaccompanied to Great Britain in what became known as the Kindertransport.

The UK was unique in having agreed to take 10,000 refugee children, and not only Jewish ones, who were in danger of persecution from the Nazis in Germany and other countries in Europe. Homes were to be found for these children – the children on the transports that started in November 1938 and ended in August 1939, just before the outbreak of war. Horst was one of those lucky 10,000. As it turned out, the move of the Seidels to the UK had been a godsend for Horst and, what is more, they helped Luise to escape also.

Chapter 4

Goodbye, Berlin

The railway station in Berlin was seething with people. Children, accompanied by one parent, sometimes two, stood in little groups awaiting the train for Hamburg. The unmistakable pounding noise and the smoke and steam of the great engine preceded the appearance of the train. There was the twelve-year-old Horst and his mother among the crowds. Stiff and nervous, like all the other mothers and fathers, Luise tried to keep the tears from her eyes. But the breaking hearts were hard to hide, and the poor young woman kept her handkerchief tightly squashed in her hand as she held the shoulder of her son. This child was her world. He was what she lived for. He had to go, to be sent away from a loving, caring home into the darkness of the world.

'I will follow you, Horst. Don't forget that I am coming to England very soon.' She looked hard into the child's eyes to make sure he was paying attention. She was like all the surrounding parents, intent on telling their children that they were loved, and that was why they were being sent away. But the boy knew what was going on. He looked at the yellow star on his mother's sleeve. He knew of the sadness that had descended on their little family already. He was trying to put the whole desperate situation behind him. Horst was projecting himself into a new life, an adventure. He knew that his mother had applied to emigrate to England, and it was almost certain that she would follow him.

Both their lives were now in a fragile state. The febrile atmosphere of children parting from their parents was being repeated that day all over Europe. And time was running out. The whole operation had to be done, and was done inside ten months. Escape was everything; the alternative could not be borne.

The noise in the station increased and the smoke and steam grew thicker as the train filled with children. Innocent and beautiful, full of excitement were the faces that beamed out of the windows of the long train. Hands were waved in goodbye, and handkerchiefs fluttered as the great beast pounded slowly out of the station.

As his mother told Gerald in later years, she returned, empty and dazed, to her father's home. She felt the shock of leaving her child, and collapsed into her father's arms, quite unable to speak. Hermann was patient. He was sad for her loss and for his loss too. The child had kept them both going each day. He waited for her recovery silently. Soon he suggested that she should sip some coffee and brandy, and as evening fell, he told of two surprises he had for her. That afternoon he had been to visit a friend of his, Herr Meyer Woolfson, a jeweller in the town.

The old man explained to the jeweller that he wanted something of value that could be hidden in Luise's clothes – a diamond ring perhaps. Both of the men knew that this would be all that the young woman would have. Only a small, token amount of money was allowed to be taken out of the country by Jews.

And so it was that a ring was chosen. It held a diamond of remarkable purity, of magnificent cut. It was a ring that cost Hermann a good deal of money. The idea was that, in an emergency, it could be redeemed in England. When the deal was done and the ring was stowed away, Hermann's old friend advised him quietly that he should apply to go to Palestine, saying that he would have no bother to get in with his money. Then it was that Hermann's voice started to rise in anger, saying that he was a businessman, and that he was not giving all his savings away with the excessively bad exchange rate on offer for his money. In a shaking voice he announced that his money was what he had worked hard for all his life. It was for his retirement. He told the jeweller that he had lost all his first fortune in the years of hyperinflation during the depression. Meyer Woolfson had to listen to this long story, which ended with the old man saying that he couldn't start up in business again, and that he was too old for the Nazis to bother with him. He was of no use to them, he thought.

That evening, Hermann talked to Luise about her future and how they would communicate with each other when she managed to get away. Their voices became muted and a sadness crept in. She had lived with her father now for years, and cared deeply about him. He and Horst were her family. But she knew she had to be firm in her determination to leave. She had to face facts.

Thoughts of the future rushed through her head. She had been accepted for training as a midwife in Oxford, but not before a medical examination in Berlin, when they had checked that she was of the required weight for the job. Luise was a slight woman and not very heavy so she contrived to put weights in her coat pockets for the weighing and got away with that ruse. Being nearly thirty-six years old, she was too old for full training as a State Registered Nurse, her first choice.

Now Hermann took out the ring. He held it up, explaining that this gift to her was something for her to fall back upon should she run out of money. He told her gruffly to sew it into her coat. Tears were not far from his eyes. She took the ring, and she also felt the tears, never far away, come rushing back to her eyes. The ring, he told her, was valuable, and any honest jeweller would recognise that. Then, turning away, he said that there was something else but since she had had such a hard day it would keep until the next day. She pleaded to be told what else he had to say. The old man shuffled over to the sideboard and produced a letter. He explained that he thought it had come from her sister-in-law, Thea. He recognised the handwriting. Luise was startled. She could not restrain herself from taking the letter. She knew that her ex-husband's sister, Thea, would not write for nothing. They had been close friends in the past. Thea had loved little Horst as a two-year-old. He would run around wildly, trying to talk and getting the words all strung together and mixed up. Many afternoons the two girls had spent walking in the park with the child, and then meeting up with Paul for coffee and chatter. Paul had also adored his son. He had photographed him endlessly and had made a beautiful album of baby pictures, a book decorated with leaves and flowers, with carefully wrought lettering in coloured inks, all done with great skill and artistry. She had loved that album and gave it to her son who has it to this day.

Luise knew that Thea must have some news of interest to her, and she pleaded with her father to let her read the letter. Shaking with emotion, she retired to her bedroom, the letter clasped in her hand.

Breslau, March, 1939.

My Dear Liesl,

It is a long time since I have written to you, and things here are not good. I am lucky in a way, in that as you know my husband's name is Haesler and that is not a Jewish name, so they are leaving me alone just now. However, sad to say, my dear brother, Paul, your ex-husband, has been arrested and put in a concentration camp. Tomorrow I go to see if I can visit him. His dance-band has broken up, and people all around are wondering which way to turn. I have one or two friends staying with me who are scared that they will be arrested. They feel safer here in my house.

One man I know spent fourteen hours on trains with his fifteen-year-old son, afraid to get off for fear of being caught by the Nazi police.

I will be in touch with you again in a few days when I return from the camp. I hope then to be able to tell you how the land lies. I hope that you and Horst are managing OK in Berlin, and your father is well.

Love from Thea.

On reading the letter, Luise pictured the scenes of those first years of happiness with Paul. Those had been her best times. Those were a few golden days, before Paul went off the rails with women friends and late nights in the dance hall.

The SS *Manhattan* rose high above the refugee children as they lined up ready to board her in Hamburg that day in March 1939. The boys, maybe about a dozen of them, had gathered in a knot to gaze up at the magnitude of the ship.

They were fascinated by the enormous size of the ship's funnels. The two large funnels were painted red with their tops in stripes of white and blue. The decks, six of them, were painted white and the hull a shiny black where it hit the blue water of the North Sea. She had provision for over a thousand tourists in cabin, tourist and third class. It was a luxurious liner, built to carry wealthy tourists to and fro across the Atlantic. But on this day, the passengers were hundreds of refugees leaving Germany, eighty-eight of them children, almost all of them travelling unaccompanied, overseen by the Save the Children organisation. Horst was in the most excited state of his young life. The parting from his mother now behind him, he decided that this was to be a great adventure. Somehow he had developed a capacity to put unpleasant thoughts and bad experiences behind him, and this skill was to stay with him for the rest of his life. However, he would later discover that they were always there when his memory was jolted.

It was a beautifully decorated and furnished ship, and except for the youngest children who seemed a bit dazed and unsure, the young refugees were thrilled with their experience. The crew were kind and welcoming to them. They each exchanged their one Deutsche Mark for English money. Horst was sick at the time of the money changing, and was lucky enough to receive a penny more than the others. They were each given fruit, drinks and sweets, and balloons were hung around their quarters. Games were organised so that it was almost a party atmosphere.

The SS *Manhattan* was bound for New York, her home port. On this voyage, the ship was scheduled to stop at Le Havre in France and Southampton in England before crossing the Atlantic.

When the ship docked at Le Havre in the early hours of the morning, the older boys, hugging the rail, were dumbfounded to see a great eagle sculpture on the quayside that looked like a Germanic symbol. For a few minutes they thought they had returned to Germany, and fear was on their faces and in their hearts. But it was found to be a false alarm, and they were soon sailing on their way to the south coast of England, and the welcome sight of the white cliffs of Dover.

Chapter 5

New Beginnings

Horst, along with several other children, was taken to a hostel on the coast in the south of England, near Margate, where, although the children were safe and looked after, they always complained of feeling hungry. The hostel must have been short of money and did its best. After a week or two the children were given pocket money. They spent hours deciding on the best way to spend their three pennies a week. The favourite purchase was chewing gum because it lasted longer.

Horst would lie in bed at night thinking back to the wonderful visits to the Kurfürstendamm, the fashionable street in the centre of Berlin, bustling with coffee houses. The wonderful aroma of the coffee drifting from every café and the sound of little orchestras playing the music of Strauss came back to the boy as he lay in his narrow bed. It had been one of his grandfather's treats for himself and his cousin, Marion, to go into a favourite café. There they had the thrill of being allowed to go to the marvellous cake counter to choose a cake for themselves. It was a lost magic.

He remembered wonderful Saturdays strolling in the Tiergarten, sometimes visiting the zoo to see the lions and tigers. How he and Marion had screamed with excitement on those sunny days with their grandfather. How was she, and how were her parents? They must be trying to leave Berlin. Marion would be fourteen now. Did she remember those times at the KaDeWe department store, where they went some Saturdays? That was their favourite outing. How he wished he could be there to see that food hall again – the colours, the smells of food, of coffee, and seafood and the fish swimming in the great tanks.

It was assumed by those responsible for the hostel that all boys of Jewish extraction should go through the ceremony of bar mitzvah whereby, on reaching the age of thirteen, they are inducted, in synagogue, into the responsibilities of adulthood. To that end, Horst and other boys were given lessons in Hebrew so that they could read the required passages during the ceremony. Horst joined in only because he felt forced to do so, never having been brought up as orthodox Jewish. He would have preferred more lessons in the English language to those in Hebrew.

Trouble came when Horst and a few of the other boys contracted chickenpox. Covered in horrible spots, they were confined to one room. There they tried to read or play games with pencil and paper, but mostly they were miserable – and hungry.

Horst had a German/English phrase book, and he found a page explaining how to ask for food in restaurants. In a terrible English accent he would call out, as if ordering food from a waiter, his desire for lamb chops or pancakes with orange. The boys would shout for him to shut up when he told them of the German translation. Although their stomachs were rumbling, Horst would carry on translating – 'cream sponge with chocolate sauce' – as pillows and slippers flew around him. When the fun and laughter had subsided, they would become quiet again, lying in their sick beds and dreaming of past days at home, and the puzzle of why they had to leave their families. Horst would think of his mother, and he would have to admit in his heart how it would be great to see her. Luise too was longing to see her son. Each night when she went to bed, her thoughts turned to her son, and she prayed to God to keep him safe.

After some weeks of tension, of packing and re-packing, of rushing to the letter box in the morning, Luise's papers came at last. With the aid of the Seidels in London, she had managed to get her papers to leave Germany. She arrived by plane in London, her English phrase book in her pocket. When her feet hit the tarmac, she looked up at the airport buildings and the blue sky above, and gave a great sigh of relief that she had made it out of Berlin. Luise was a strong woman, but beneath her calm exterior she felt shaky and unsure of herself. Her knowledge of the English language was very slight. She had, in the past weeks, spent hours trying to remember useful phrases. She hoped she would be able to make herself understood. Luckily, her old friends from Berlin were at the airport to meet her. There were smiles and just a few tears as they clasped hands, joy and thankfulness filling their hearts. She spent a relaxing few days with them. Her chief thought was that her son was not far away. She mused on her position with hope for a future life. She had in her purse exactly ten pounds, allowed by the authorities when she left the country, and in the third button of her tweed coat she had sewn the diamond ring given to her by her father.

At last came the time for her to travel to Margate from London to see Horst. This moment had become what she was living for – to see her beloved boy again. As she stood in the railway station with Rudi and Herta Seidel and young Hardy, she struggled to contain her excitement. She was so glad that she had been able to buy her train ticket for the journey from London to Margate. She boarded the train after their goodbyes, and as she gazed at the landscape, she dreamed of her future.

Arriving in Margate, she got to the door of the hostel by using her phrase book and haltingly asking passers-by to show her the way. When she was reunited with Horst, there were hugs and tears, so much so that the two of them drew the attention of other boys and staff, who were much moved by the scene. Being past twelve years old, Horst decided he could not show as much emotion as his mother. Gradually, he was growing up, and feeling a bit of independence, and he tried to hide his great feeling of love for his mother.

Luise was given some lunch at the hostel, and as it was a fine spring day they set off for a walk along the promenade by the seashore. They spoke in German, of course, as neither of them had yet learned enough English. She told him that his grandfather had missed him terribly, even if he had tried not to show it when they said goodbye. The boy was sad when reminded of the great bond he had had with his kind old grandfather.

They talked for a while about the bad situation at home. Horst discovered that the old man had moved from Bamberger Strasse to an old people's home on the other side of Berlin. His mother explained that this was the only thing they could do since she had to leave Germany and he could not look after himself. Her sister, Erna, was also unable to help, even though still notionally in Berlin with her husband, Erich. They were constantly on the move to try to evade the Nazis who wanted to take Erich to a concentration camp. Neighbours had told them of such a 'call' when, fortuitously, Erna and Erich were out shopping. Erna, Erich and their daughter, Marion, had an affidavit from Erich's cousins in Baltimore to emigrate to the USA. With that expectation they had already sent some of their furniture to Baltimore. But entry to the USA was controlled by a quota and their number had not yet come up.

Horst's mother was ashamed, she said, that she had not written to tell him all of this, but she had been so busy trying to get her Papa settled and prepare for her own departure to England.

Horst took in all this information. He asked about all his grandfather's furniture and gramophone and was sad learn that it had been sold for next to nothing. He also learned that the landlord had refused to refund them the two months' rent that had been paid in advance.

Luise was allowed to stay for a meal at the hostel before taking the train back to London. She got the boy's attention, saying pointedly,

'Now, Horst, tomorrow I go to Oxford to start my training as a midwife. I promise that, as soon as I can, I will try to get you to come to Oxford. I'm sure there must be a refugee committee that I can ask.'

And so their first reunion ended, this time Horst waving his mother goodbye as the train steamed out of Margate Station. And this time there were no tears.

At every opportunity, in her solitude, Luise would try to learn to pronounce and learn some useful phrases. She struggled to control her apprehension that her poor English would let her down. Soon she found herself boarding a bus for the Radcliffe Infirmary in Oxford. As the bus passed by some of the fine college buildings, she was bemused by their beauty and the apparently slow pace of life in the town. In those last days before the war, who knew of the chaos that was to burst upon their world a few months later?

The Radcliffe was a well-run, efficient complex where nurses and matrons were smartly dressed in starched caps and aprons. They went about their ministering to patients with a sense of duty and care.

Luise was welcomed gladly, and soon she was installed in her own room in the nurses' quarters, and with her books she was able to study seriously to gain her qualifications. From now on she would only be 'Luise' or 'Nurse Wiener' and not called by the affectionate 'Liesl'. It was a luckier position than she had expected to find in England. But the times were frenetic; men and women were being called up for the armed forces. Everyone expected war to come soon, and Luise was grateful to the staff for making her feel needed in this time of crisis.

It was the month of May 1939, and Luise lost no time in contacting the local refugee committee. She asked them to try to have her son Horst transferred to somewhere nearby, so that she could see him when she was off duty. The next Sunday, an announcement was made by the vicar, Dick Milford, in St Mary's, the university church, appealing for anyone willing take a refugee boy, thirteen years old, whose mother was training as a midwife at the hospital. 'We are looking for some kind person, or a family who could answer this appeal in these dreadful, unsettled times.' The call was answered by two sisters, elderly unmarried ladies, Eleanor and Monica Ewbank. They thought they could manage to board the child. And so it came about that Horst arrived to live in Oxford.

War intervened to change the beginnings of a happy time with the Ewbanks. Like many ladies of their advanced age and genteel upbringing, they felt impelled to do voluntary work in London to help the war effort. The ladies sent him to the Christ Church Cathedral Choir School, which had been set up to teach the boys of the cathedral choir. The cathedral was in effect the magnificent chapel of Christ Church College. The Bishop of Oxford had made it his seat, but it was never called Oxford Cathedral.

The choirboys at the school were all boarders, and for a period Horst, though not a choirboy, joined them as such, when the Ewbank ladies went off to London. As well as boarders there were also day pupils. At this school Horst was happy, and felt at home. It had been chosen for being small, so that a boy with no English initially would not be overwhelmed. There were less than twenty boys in the class and they were full of fun. They made it their challenge to teach him English, and when lessons were done, they sat around after school pointing at things.

'Look, Horst,' one would say gleefully, '"table" or "desk".' Another at dinner would point at his plate and say, 'po-ta-toes', and Horst would repeat 'po-ta-toes' and everyone would laugh. Colloquial English was picked up very quickly by the boy, with a posh Oxford accent to boot. One fly in the ointment occurred on the very first day, as Gerald later recounted to his wife. A rather arrogant housemaster asked the refugee, 'What is your name, boy?'

Just about understanding the question, he replied, 'Horst!'

'Do you mind if I call you Horse-face?'

'No,' said the boy, with no idea what it meant.

And from that time on at that school, he was called Horse-face. He didn't understand that the name was pejorative, and the other boys took it as a fun name and liked him just the same. When they went to Boy Scouts' camp, and Horst was the cook, they sang to him: 'Horse-face, you make a darn good stew!'

'Thank you! And the same to you!' was his response.

One of the teachers at the school suggested that Horst should live with her parents in Kennington, a village some three or four miles from Oxford, a twenty-minute bicycle ride from the school. Boarding in the school would be too expensive a long-term option for the Ewbanks. It was in Kennington that Horst first acquired a love of gardening, and of growing vegetables in particular. He also assisted the neighbours in building a large concrete air-raid shelter to which many of them would go during the air raids of the 1940s. The village was on the path of the German bombers flying to Coventry and other industrial towns. The drone of the planes' engines, the sight of the waves of bombers in the sky at dusk, and the flashing of the searchlights from the ground to seek out the planes made for an eerie experience.

At the age of fourteen, Horst transferred to the City of Oxford High School for Boys. Henceforth, he was called Gerald, his middle name. Fluent in English, he left school at sixteen with a reasonably good school certificate.

Outside school, he had been active as a messenger boy for the Air Raid Precautions (ARP) local group. He was given a bicycle, a uniform and a special gas mask – these made him feel very important. Falling off his bike once, he was put in the ARP logbook as 'injured while on duty'. But this was a terrible time for the south of England and the Midlands. It was a time of massive air raids and often waves of German planes could be heard flying overhead on their bombing missions.

The question of what kind of work he would find arose. He fancied an apprenticeship to electrical engineering. This was to be at a large plant in Birmingham well known for apprenticeships combined with study. But this scheme was suspended because of war work. Then he wanted to be a chef. This wasn't possible either. And so with the help of some friends, whom his mother had come to know locally, he was taken on for farm work. These good friends turned out to be a great influence on Gerald's life, as did the choice of working with animals.

CHAPTER 6

The Spooners of Oxford

Luise was settling into her nursing job and orientating herself with Oxford and her new environment. Mostly her thoughts were with her young son, whom she occasionally still referred to as Horst, as he had been called in Berlin. Gerald no longer wanted that name from his past and it upset him when she used it.

Luise had heard about the Spooners at 9 Polstead Road, in north Oxford. She had met up with a few refugees from Germany who had told her about the Sunday open house that was held there by these very kind and gentle ladies. To meet Ruth and Rosemary Spooner was a privilege and an experience. The women were cousins, and neither had married. They were at the disposal of all who called on them for assistance. Ruth, a wonderful pianist, played each Sunday for the assembled displaced persons who arrived at that famed address in Oxford. Generosity and kindness were limitless for Rosemary and Ruth. Their home was indeed a light in a dark world.

Luise started going along to these Sunday get-togethers in Polstead Road. She was told of how Rosemary's father, when he was the warden of New College in Oxford, was famous for his habit of mixing up his words. For example, he would say, 'Is the bean dizzy?' for 'Is the dean busy?' or perhaps, 'You have hissed my mystery lecture' for 'You have missed my history lecture'. These humorous slips of the tongue, which became known as

Spoonerisms, meant little to Luise at that time, but she laughed anyway.

At first she was a bit in awe of these two tall, elegant ladies. It was clear that they had been born to higher things than making tea and cooking meals for refugees. They were both graceful and kind, fine examples of true Christian behaviour. They were well read in political theory, and in their everyday lives they tried to follow the teachings of Christ, that is to 'love thy neighbour as thyself'. However, they did not discuss religion, nor did they try to convert anyone. Luise felt the warmth of their welcome, and could relax in the break from the strict routine at the hospital.

And she could sometimes speak German with some of the other guests. They were interested in her story, and also in her work at the hospital. Gradually Luise's English conversation improved, and she was able to impart to Rosemary and Ruth her fears and worries as a woman cut off from her family still living in Berlin. Of course, she told them of her son, and of how well he was doing at school and of how he was living with an elderly couple in a village some distance out of Oxford. They suggested she bring him along.

The next time Luise saw Gerald, she told him all about the Spooners. Luise was an excitable, emotional lady, with continental mannerisms. When she spoke, her eyes lit up, and her shoulders were raised and lowered as she added impetus to her story. Gerald was impressed and the day soon came when they were to visit the house in Polstead Road. Gerald remembers how they took special care of their appearance. He was warned to have good manners, and to be attentive when spoken to. As a result, some of his mother's excitement rubbed off on him, and as time would tell it was truly a significant meeting for the boy.

Along the roads of that beautiful town they walked past detached and semi-detached three-storied Victorian houses that seemed quite grand, the small gardens beautiful in the bloom of spring. The calm of the street and the very few cars on the road on that peaceful

Sunday, did not prepare the fifteen-year-old boy for what lay inside the suburban house. Here his ears were assailed by the sound of the grand piano. When they had been welcomed and were seated in that packed sitting room, they relaxed together, feeling almost at home.

The guests were an unusual crew: Polish airmen, French soldiers, Austrians, Czechs and Jewish refugees from Hitler's Germany were all accommodated in that room, in various positions of comfort, on the chairs and on the carpet. The boy was mesmerised by the atmosphere, the music and the soft air of attention that was being paid to the beautiful pianist, Ruth. She was slim and graceful, her head inclined slightly, her hands lifting gently over the keys of the piano. She finished with a smile to her audience, acknowledging the applause. She was a highly accomplished pianist often invited to give recitals for charity in different parts of the country. She practised the piano diligently for three hours each day – though not on the grand piano but in her music room upstairs in the house.

Rosemary, unlike her cousin, was of a more commanding presence, a town councillor and member of many committees. But she was every bit as kind. Often Rosemary would accompany Ruth on the cello, adding to the variety of the music. This day, however, she emerged from the kitchen pushing a tea trolley holding two large teapots and plates of sandwiches and cakes. These two cousins had been brought up in large houses with many servants, though now they had none. They were connected to leaders of academia, to politicians and to leaders of the Church of England (the Archbishop of Canterbury, William Temple, was a first cousin). They could also trace their family back to William Wilberforce, a leader for the abolition of slavery. Yet they had thrown this privileged way of life aside to help the less fortunate in this time of upheaval and of war. Also, it has to be said, they gave support to many charities. They had a special sympathy for the totally deaf, who they felt missed out on many of the joys experienced by those with hearing. Both cousins had mastered the sign language that would allow them at least to communicate with the deaf.

Gerald was enchanted. He smiled to anyone who spoke to him, and when it came time to go home, he was sad to have to leave these interesting, friendly people.

As they were leaving, Luise shook hands with both cousins and with some emotion told them how pleasant they had made her stay in Oxford. She was about to leave to start nursing training in London after finishing her midwifery in Oxford. She thanked the Spooner ladies profusely. They in turn complimented her on her well-mannered son, Gerald. This gave joy to the lonely Luise, who added that she was proud of him and his performance at school, especially his quick mastery of English and his high marks in mathematics.

It was then that Ruth Spooner threw the bombshell, telling Luise that Gerald could come to stay with them. They could fit him in and he would not have so far to cycle to school every day, and he would have more time to study for his exams. The Oxford High School for Boys, Ruth concluded, was only a five-minute bike ride away. Gerald's eyes opened wide at this. He could not believe his luck. And so, within weeks he became a ward of these two benefactresses.

They made him part of the family. Gerald would help his 'aunts', as they insisted he think of them, when they were entertaining, carrying food from the kitchen to the dining room, assisting with the washing up of the dishes and going on errands. On Saturday mornings he would queue, along with many other people, for an allocation of cakes, a luxury in those days of war. The cakes were, of course, for the tea of the Sunday afternoon open house. Even his childish peccadilloes, such as stealing one or two dried dates here and there from a huge pottery jar in the pantry, were never commented on. To this day he feels guilt for depriving the household by pilfering some of their boxed dates, which could not be replaced during those days of food rationing.

For much of that year Gerald cycled back and forth each day to school. In the evenings he had dinner with the ladies, and whoever else they were entertaining at the time. Gerald absorbed the gentle manners and love of culture, music and books of his adopted aunts, and, importantly, the example set by their way of life. They were people brought up with a religion that stressed the love of one's fellow man, no matter what their circumstances, colour or religion. Indeed, there was one particular sitting at lunch when a bishop and the local dustman, both friends of the Spooners, and some African students sat together, all holding conversations. Sometimes, Ruth would lift a book by one of the Brontës, or more often by Charles Dickens, and proceed to read to the assembled people at the table. Gerald found these almost professional performances by Ruth to be exciting and moving, for she had learned from her mother to read in the fashion of Dickens himself; that is, in the exaggerated dramatic style the great man had used to get his story across to his audience. Ruth's mother had actually been at some of Dickens' readings and had picked up his mannerisms and style and taught them to her daughters.

For the boy this was the start of a friendship that lasted into his later life. Ruth even travelled to Edinburgh to be at Gerald's wedding to his first wife, Sheila, in 1952. They gave the couple a generous wedding present of one hundred pounds. With this they bought a posh dining-room suite.

Some years after that, both Rosemary and Ruth came up to Scotland on holiday. Gerald had the great pleasure of accompanying these two stalwart ladies on a walk in the Pentland Hills. When, later still, his two children were born, the two ladies sent cards and gifts; truly they were like close relatives to Gerald. The friendship continued until both of these kind women reached a grand old age. Gerald felt that knowing the Spooners had been a great blessing, which had influenced the rest of his life in many important respects.

The Spooners' giving of themselves extended to their possessions, many of which were bequeathed to others even during their lifetime. Some of their treasured family silver cutlery was bequeathed to Gerald, as well as a watercolour painting by the Victorian painter Aubrey Waterfield, which had been given by the artist to his friend Dr Spooner. The painting then passed to his daughter, Rosemary. That painting had hung above Ruth's desk, from where she conducted her copious handwritten correspondence with her many friends and acquaintances.

Many of these loving letters from both Ruth and Rosemary were sent to Gerald, always addressing him affectionately as their adopted nephew. He treasures the letters to this day. The Waterfield painting has pride of place among the other pictures in Gerald's home. Among the many bequests made by the two cousins was their house, 9 Polstead Road, which they left to Ruskin College for use by married overseas students. Subsequent to Ruth's death at the age of eighty-seven, the Principal of Ruskin College, H. D. Hughes, wrote of Ruth Spooner in an obituary column, 'I regarded her as one of the few human beings who were saints.' The same could have been said of Rosemary Spooner who died some years later.

Chapter 7

Farming

The year of 1942 was a hard one for most of the people of Britain. Life was disjointed, and emotions were running high. Young men were called up for the army, bombs fell on the cities each night, and generally there was a feeling of displacement and tumult in the environment. In the countryside, life was safer, but it was still a cold, hard, laborious lifestyle for most.

In later life, Gerald took some time to talk to his family about his life in England when he reached the age of leaving school. Nothing was easy, either in industry or in education. Unable, because of the war, to get the apprenticeship he had wanted, a job was found on a local farm for the sixteen-year-old. Here he was unpaid, as a 'trainee', but received his bed and board in exchange for farm work. Hand milking fifteen cows at five o'clock in the morning was a great trial for a town boy who had never before been close to a cow. He fell into bed after evening milking, totally exhausted. This was much to the amusement and disdain of the head cowman and his wife with whom he lodged.

He moved to a better situation on the farm of Mr and Mrs Rowe. He was paid the five shillings that were left over after Mrs Rowe had taken his board money. They were a kind couple and treated him like one of the family. He shared a room with their son, who was a few years older than Gerald, but they got on well. The work was varied: looking after pigs and sheep, working with horses, and helping with the field work. These were all great experiences for the young man and he felt happy at the thought that this could be the start of his career.

The farmer and his wife were members of the Congregational Church. This denomination of the Christian religion stressed personal responsibility and it appealed to the young man. Later in his life, Gerald was baptised and became a Christian and joined that denomination. It is indeed a wonder how this seventeen-year-old boy found his way through the plethora of faiths that were thrown at him. First he encountered the Christ Church Cathedral Choir School and frequent attendance at high Anglican services, which to the young refugee boy seemed devoid of feeling though not of ritual. Then he was contacted by a Jewish family living in Oxford. A Professor Cecil Roth, no known relation of the famous American novelist Philip Roth, contacted him for the reason, as they saw it, that he should attend their synagogue on Saturdays. Gerald did not want to do this, because he was already attending lessons at the high school on Saturdays, necessitated by wartime conditions. Against his will, he was made to go to synagogue, until he eventually rebelled. He was even to be a bit player in the Passover ceremony at the Roths' home. He played the youngster who asks the ritual questions that are then answered by the head of the household as part of the retelling of the story of the exodus of Jews from Egypt. Everything was in Hebrew, and to Gerald this was ridiculous as he did not understand a word of it. He had not been made to attend synagogue or to be so religious in Germany. He was sure in later life that the exemplary Christian life of Ruth and Rosemary Spooner had played a large part in his journey of faith.

At this time on the farm, Gerald was also a member of the wartime Home Guard. Because he was considered intelligent and educated, he was allowed to man the phone in the little headquarters shed. The other farm workers, burly men all, defended the neighbourhood in all weathers, armed only with large homemade clubs through which they had driven huge nails. Uniforms and proper arms came later. One night, on a Home Guard exercise, while Gerald was sitting snug beside his phone, Phil, the farmer's son, returned covered in cow shit, having crawled through a field with others on a training exercise. He smelled to high heaven. 'I suppose you think this is funny,' he said to the boy manning the phone, who was trying not to laugh.

Sometimes Gerald accompanied the Rowe family to the Congregational church in Luton. It was there that the minister, the Rev. Sansom, preached. He had become a friend of Gerald's mother while visiting hospitals where she worked and it was he who had introduced Gerald to the Rowe's excellent farm, where he did indeed learn much about farming, which was to stand him in good stead later. William Sansom and his wife became lifelong friends to Gerald.

More hard and responsible work followed when Gerald was asked to help out on a farm where the farmer had died. His widow, Mrs Bailey, was a good friend of the Rowe family. Mrs Bailey's foreman had been run over while he was walking home drunk in the middle of the road. The only other worker on the farm was a simpleton guy who would not come to work if the foreman was not there. So she was left with only her eighteen-year-old son to run the place. Three months of particularly hard work ensued for Gerald without even one half day off. While Mrs Bailey's son dealt with the field work, tilling the land, sowing seed and spreading fertiliser, Gerald was left to look after fifteen dairy cows. Fortunately he was good at hand milking by then. In addition there was other stock on the farm to be fed and looked after: chickens, a few pigs and a bull in a pen that frightened him. The bull was accustomed to being looked after by the now absent 'simpleton' and clearly resented the new worker. At nights on this farm he could hear the sound of wolves howling in Dunstable Zoo, less than a mile away, so clearly that it was quite scary.

The Spooners then took another hand in Gerald's life and told his mother that they thought Gerald should go to university to study agriculture – they felt he was too clever to remain just in farm work. Luise of course agreed, full of pride that her son was thought of so highly. None of her family had ever had a higher education. Through university contacts it was arranged for Gerald to take up work at the animal research farm on Huntingdon Road on the outskirts of Cambridge. This would provide a better place for him from which to apply for study at university and to prepare for it.

Gerald lived in a hostel in Cambridge occupied by young men and young women, all refugees, similar to him, and all working or studying. He really enjoyed life in the hostel with these young people. They were a bright and intelligent bunch and Gerald became friends with several of them. In the evenings they often used to sit around talking about what they were doing and of their hopes to do great things with their lives.

One thing the hostel was not very good at was food. Gerald put it down to the difficulties the cook had in providing sufficient meals with the wartime rations available to her. But perhaps that was not the reason, as things improved at weekends when the warden of the hostel, a Miss Souhami, took over the catering. Most evenings during the week, many of the boys and girls could be seen in the kitchen scratching up food from wherever they could find it – mostly bread with some spread on it. One of the boys was alleged to have got hold of a seven-pound tin of peanut butter, but Gerald never saw the tin and was sceptical about its size. Gerald himself was guilty of gluttony on one occasion. He was working in the poultry section of a research farm and the workers there were allowed to take home cracked eggs – usually not more than one a day – that could not go to the official collection centre. Gerald accumulated ten such eggs, keeping them in his bedside locker, and one evening he made a giant omelette from the lot. He was unable to face an omelette for a long time thereafter.

Often in the evenings, the boys and girls in the hostel would sit in their common room listening to music on a record player and sometimes engaging in philosophical discussions, which were always hijacked by one of their number, a philosophy student, with his favourite phrase 'it all depends on what you mean by . . .' On one of these evenings, Gerald was interrogated about his work on farms and his present job. This, they assumed, was exciting involvement in research under the eminent John Hammond, a world-renowned animal physiologist.

Gerald had to disillusion them, admitting that he was at the bottom end of the pack, mostly engaged in cleaning out henhouses and carting animal feed around the farm. Gerald's boss was Michael Pease, a geneticist who was creating breeds of chicken in which it was possible to separate visually the male chicks from the female chicks; these were known as auto-sexing breeds. This was a useful genetic challenge as the males of egg-laying strains are not profitable to rear. It was a clever idea but in the end was not commercially successful as the strains of chicken involved were not the best at egg laying and Japanese experts had developed a means of telling the two sexes apart at a day old.

Gerald was, however, indirectly involved with John Hammond. Dr Hammond had crossbred, by artificial insemination, a great Shire mare with a little Shetland pony stallion. He had also produced the same crossbreed by insemination of a Shetland pony mare with semen from a Shire stallion. His purpose was to find out if the uterine environment of the two mares would affect the ultimate size of their crossbred foals. Although carried out with only two pregnancies, the results were sufficiently dramatic to become part of animal breeding legend. The Shire mare's crossbred foal, when fully grown, was twice the size of that from the Shetland pony mare. And the larger foal was also much more placid, as Gerald found when he harnessed these horses to a cart for his rounds carrying feed for the animals on the farm. The boys in the hostel made some fun of this story, saying they would have to remember this result when looking for a wife.

When the girls had left the common room one night, the boys asked Gerald why he never invited any of the girls to go out with him in the evenings, as they thought he was quite liked by them. Gerald was somewhat embarrassed by this turn of events. He had to admit to being rather shy of girls and perhaps sexually retarded. He blamed his mother for that as she had often asked him not to turn out like his father – a serial womaniser. He admitted to his friends that he had not taken up the offer of one of the girls to scrub his back in the bath. And another, to whom he had plucked up courage to say she was very beautiful, had responded that she was even more beautiful without clothes on. His only response had been to laugh. The other boys thought him a fool and secretly he agreed with them.

The future was still uncertain in Gerald's mind. He knew that he was expected to go to university, but he also hated the country of his birth for what the Nazis had done to him, his family and countless others. So he decided to enrol for the armed forces. He went to the RAF recruitment office to volunteer. After questioning him, the officer behind the desk said, as Gerald remembers to this day, 'Don't be a bloody fool! If you have a chance to go to university, you should go, my son!' That retort, Gerald thought, could only have been made in a university town like Cambridge. He did not sign up.

The person most instrumental in assisting Gerald with applications to universities was Greta Burkill, the chairperson of the local refugee committee. The hostel was formally under her committee's charge, and the committee had been much involved also with refugee children who, like Gerald, had come to the UK on the Kindertransport. Gerald discovered in later years, through research of the archives by Mike Levy, that in commending Gerald for entry to university she had said that he was a good and studious young man but 'unlikely to set the Thames on fire'. In later years she was to revise her view by including Gerald's name among a few from 'her' hostel who had distinguished themselves in their career.

The Spooner cousins, intent that their refugee ward should go to university, set the wheels in motion for the financing of that exercise. They donated a generous amount themselves, and applied for him to the Save the Children Fund and to a Jewish charity, B'nai B'rith, to help with the finance, which they did. Luise sold her Persian lamb coat and raised seventy-five pounds, which was enough to keep him at his lodgings for a year during the term time of the academic year. During vacations he would have to work to earn money to keep himself. University fees were paid by the government in those days.

Chapter 8

Student Days

The die was cast. It was 1944 and after some consideration Gerald had chosen the University of Edinburgh to be the place where he would study agriculture. It was an enlightened choice. Agriculture at the university had a fine history and its resources expanded greatly around the time that Gerald arrived there.

The university boasted the establishment of the first Chair of Agriculture in the UK in 1790. The department, when Gerald first arrived on the scene, was still situated on the beautiful George Square in Edinburgh, with its lovely gardens in the centre and surrounded by grey sandstone buildings mostly dating to the late Georgian period. The College of Agriculture, established in 1901, was but a few doors away. In 1947, the last year of Gerald's degree course, the university would purchase the 3,000-acre Bush Estate south of Edinburgh and develop on it a campus housing some twelve research, advisory and teaching institutions, including the field laboratories of the Animal Breeding Research Organisation (ABRO) at Dryden, where Gerald was to work in later years, and ABRO's Mountmarle pig station, both adjacent to the village of Roslin.

On the day Gerald arrived in the northern city it was clear and sunny. The month was September, and, looking back, he recalls how pure and fresh the air felt. As he strolled along Princes Street, the charm of the city started to work on him. He loved the drama of the castle above and the lovely flower gardens below.

Lodgings had been found for him by the refugee committee in Oxford. This was one room in a top-floor flat on Gladstone Terrace, just a short walk through the Meadows from the university. The landlords were a Mr and Mrs Young. Gerald received two pounds per week from the money donated and administered by the Save the Children Fund. From this amount he had to give thirty-two shillings and sixpence to Mrs Young, leaving him with seven shillings and sixpence for everything else, including his lunches and travel (one penny on the tram). He had breakfast and an evening meal from Mrs Young, but got rather tired of the frequent mutton pies with mashed potatoes, or mutton pies with baked beans. Also resident in the apartment was a large Airedale dog. Sometimes he would give half his dinner to the dog because if he left anything on his plate he was given less to eat the next day.

To get to his lectures Gerald had to travel by bicycle or tram to and fro across Edinburgh. In his first year he had physics lectures in the centre of town, had to travel to King's Buildings in the south of the town for zoology and chemistry, and north to the Botanic Gardens for botany. That was a lot of travel but to him it was a pleasure. He found it enlivening to see the city in this way.

Some wit had described the agriculture degree course in Edinburgh as the best possible training for a career in journalism. Over its three years no less than seventeen different subjects were covered, or at least introduced. As the agriculture class started off with only twenty students – sixteen young men and four girls – many of the classes were taught jointly with the pure science or veterinary students, apart from those topics only relevant to agriculture. There were subjects as diverse as political economy (Gerald always wondered who had decided on that subject as a 'must' for agriculture students), mycology, geology, engineering fieldwork and forestry – to mention just a few.

Lectures and the lecturers were varied in style and apparent competence. At the bottom end Gerald recalls political economy, where the lectures consisted solely of the lecturer reading on each occasion a chapter of the book written on the subject by her erstwhile professor. Had it not been for the fact that students had to place attendance cards in a box at each lecture it would have been simpler merely to buy and read that book. Two other 'lows' stuck in Gerald's mind. One was in a course entitled Veterinary Hygiene, taken jointly with veterinary students. Most lecturers encouraged genuine questions but this particular man clearly did not. One of the class, an ex-serviceman invalided out of the army and older than the rest, and Gerald, the two most interested in delving below the surface, were told in no uncertain terms that they would be barred from the course if they asked more questions. Clearly, this lecturer had his script set out in full and did not wish to deviate.

The last such memory is sad rather than accusatory. On occasion during the forestry course the students would be addressed by the professor of the department himself. He, however, was already in his eighties – appointed in the days before a retirement age had been introduced. From time to time during his discourse he would forget what he was saying and repeat the last of his words over and over again until his memory came back. Youngsters though the students were, they knew that perhaps someday they too might come to this state and no one ever joked about the professor's failing memory.

Of course there were high points too. Professor James Ritchie and Dr Willie Gross enthused Gerald with zoology, Charlotte Auerbach with the foundations for his interest in genetics, and Hugh Donald with genetics applied to animal breeding. Botany was a joy, not for the brilliance of the lectures but for the infectious enthusiasm of the lecturer who, magnifying glass in hand, would study each minute part of a plant or its flower as though it was the first time he had ever done it – yet he must have been doing it for the past thirty years.

Tony More, the Reader in Agriculture and the Director of Studies for the whole class, was a brilliant and dedicated man, an inspiring teacher to listen to on his own subject of animal husbandry. He was co-author, with James Scott Watson, of what was for several decades *the* standard textbook in Agriculture – *Agriculture: The Science and Practice of British Farming*. The book went through eleven editions. Sadly Tony More died in 1947, the last year of Gerald's course. He was only in his fifties but had been in poor health for much of his life, having developed health problems in the First World War. For that reason Tony More had had to turn down offers of even more prestigious posts.

His widow, Alison More, became a lifelong friend of Gerald and of both his first and second wife. She was also godmother to his two children. Alison herself had a remarkable intellect and knowledge, the equal of the many eminent people from the university and elsewhere who were among her large circle of friends. She had shared, as a student, her future husband's training and experience in farming and supported him in that. She also had a particular love of gardening. It is doubtful that there was a plant that she could not identify. In later years that was a help to Gerald and his wife when they established a garden. But Alison had three tragedies in her life. The first was when she lost her husband, who continued to live on in her thoughts and her conversation ('Tony would have thought that') for all her days. Then her son, Brian, a doctor by profession and also friend of Gerald, died in a freak accident whilst swimming in the Red Sea. And not long thereafter, her daughter, Joan, a beautiful girl married to a farmer, died also. That Alison overcame these tragedies and continued as a caring and active person is remarkable and a tribute to her inner strength. She got much joy from her grandchildren and later her great-grandchildren. When she died, aged ninety-nine, the family gave Gerald the honour of paying the tribute to Alison at her funeral.

Gerald did well in every class apart from physics. Here he had problems. A kindly Professor Kemp-Smith, to whom he had been introduced by friends, paid for extra tuition in this subject. So through that help, and reading a school textbook the day before the final exam, he was able to pass the course.

Wednesday evenings were spent at meetings of the Student Agricultural Society, which Gerald joined in 1944 during his first year and became increasingly involved with. Its members were from both the university and the College of Agriculture. They joined in discussions and sometimes in debates with outside groups such as the Young Farmers' Club and, of course, there was the occasional social. Gerald contributed actively and eventually he became its secretary. Soon a Scottish Association of Agricultural Students was formed to represent agricultural students from all of Scotland, and, to Gerald's amazement, he was elected its first president. At the first joint meeting in Aberdeen, the students got him drunk for the first, and probably last, time in his life.

At these meetings, Gerald struck up a friendship with the chairman of the group, Ewan Stewart. He was invited to the Stewart house in Polwarth Terrace, where Ewan's mother welcomed him with her motherly attitude. Here he was delighted to spend time enjoying homemade scones and cakes and listening to the tales of Ewan's father. The house was large and comfortable, a detached stone building in a middle-class area where all was calm and refined. It was a close family life, which Gerald loved. Mr Stewart was a retired civil servant concerned with land allocation and disputes in crofting areas in the north of Scotland. He listened constantly to the news bulletins following the progress of the war in Europe. This habit continued even after the war had finished.

Gerald vividly recalls the day when it was announced that the war in Europe had ended. The date was the 8 May 1945. He was still in his first year at university and he walked by himself to see the great celebrations along Princes Street and the Mound, where he watched the wild dancing of the people there and the music and cheering on all sides. He felt the warmth of the smiling faces, and in his heart he joined in the celebrations and was glad that the great and horrible bombing and killing had come to an end, at least in Europe.

As the months and years passed, Gerald grew to appreciate Edinburgh increasingly, though he had loved it almost from the start. He felt he belonged to the city and happily worked hard at his studies. It was only on Saturday nights that studying stopped. Those nights were reserved for dancing at the regular hop in the Students' Union. Gerald had taken ballroom dancing lessons, and throughout the months of the classes, he had forged quite good friendships with several girls who were his partners in the classes. But his inhibitions held him back from any real closeness. He blamed his shyness on his mother's constant warnings that he should not turn out like his father, that philanderer and womaniser.

Eventually, he did meet a girl called Jean with whom he fell in love. They were sweethearts for a couple of years, but often did not see each other for months due to their work commitments. In the end, their friendship faded and Jean met someone else. Gerald was left to pick up the pieces of his broken heart. This took a while, as he said she was the love of his life at that time.

In 1947, when he reached his twenty-first birthday, he was allowed to apply to become a naturalised British citizen, the thing he desired above everything. As part of the application process he had to advertise his intention in a local newspaper. Mrs Young, his landlady at that time, took great exception to the advertisement – perhaps she had kept it quiet from her neighbours that she had a 'jerry' staying in her house, but Gerald had no cause to ask her permission. He would have moved from his lodgings rather than forego the chance of British citizenship.

So in 1947 Gerald left his original digs in Gladstone Terrace with the Youngs. He was happy to do this as he was tired of the isolation in his room there. Through Ewan Stewart's family he was able to find lodgings with Mr and Mrs Milner in Merchiston Crescent, in the Bruntsfield area of Edinburgh. There he lived happily with the family. There was one son in the family who spent time unsuccessfully trying to teach Gerald to play golf. The main leisure interest of the Milners was playing bridge. They played contract bridge several times a week, either in clubs or with friends. After each game, they would arrive home still arguing heatedly about this or that trick, accusing each other of unforgivable mistakes. Gerald was so put off bridge by this experience that he did not take up the game until he was in his seventies.

During this time Gerald started to attend worship at the local Congregational church. This was only five minutes of brisk walking from where he was now living. He also joined the Youth Fellowship at this church. They had a weekly get-together and at the opening of each meeting a prayer was said, led by a different member of the group, in rotation. When it came to Gerald's turn he discovered that, in spite of having worried at the prospect, he could do it easily. It came readily to him, and he found the experience rewarding. As he remarked when relating these events, 'I felt that I was getting closer to embracing Christianity.'

Gerald was a regular attendee at the Morningside Congregational Church, and greatly admired the minister there, Mr Newsham, a brilliant preacher in Gerald's eyes. He was also a conscientious pastor and when the Milners needed Gerald's room he found new lodgings for him with members of his congregation in Morningside Road. Wanting to explore and encourage Gerald's spiritual journey, Mr Newsham invited Gerald to his house to discuss his developing faith with him. They had a number of such discussions and the minister was so pleased with Gerald's understanding and insights into Christian life that he expressed his wish for Gerald to be baptised without further ado and, almost immediately afterwards, to be admitted to membership of the church.

Before taking this serious step, Gerald felt that he should discuss the matter with Luise, his mother, on one of his visits to her. She herself quite happily attended hospital services led by Christian ministers, and unshakeably believed in God. However, in spite of the fact that she had never been near a synagogue since she had left Berlin, she almost threw a fit when Gerald spoke to her of his intentions. 'How could you? How could you?' she cried. 'Give up your Jewishness, and your great forebears?' Gerald was shaken and for the time being abandoned the idea. Mr Newsham was very disappointed, as was Gerald. It would be another four years before he took those formal steps. By then Mr Newsham had been called to serve a large congregation in the United States but on returning to Edinburgh on a visit he was glad to discover that their evening discussions had borne fruit.

Gerald also consulted his mother on another important issue. Around four years after the end of the war, he received a message from his father, Paul, through the Red Cross. The message explained that Paul was in San Francisco and asked Gerald to contact him. Luise had never ceased to feel bitter about her former husband and thought that he probably just wanted money as, according to her, he had always done. She persuaded her son to write to the Red Cross to say that he did not want to make contact with his father. He acceded to his mother's wishes, though in his heart he regretted it deeply.

Gerald worked in the Post Office over Christmas to earn a little money. In the summer, he worked at potato inspecting and for several years teaching the inspectors' training course. In later years, potato roguing earned him the money he needed to pay his way, especially when his salary from his regular work was very meagre. Potato roguing involves the removal of any plants found to be diseased or of the wrong variety, the 'rogues' in the crop, from fields of potatoes being grown for seed. It paid him, and his good friend Ewan Stewart, one pound an acre. By working hard and over long hours they could earn more money from this than from other holiday jobs. Ewan, who was studying law at the time, had, on one occasion, to leave the farm where they were working to attend a court case. Gerald had to finish the job on his own at double speed and to drive Ewan's old pick-up truck back to Edinburgh. As it happened, the truck was leaking carbon monoxide into the driver's cabin. Gerald got dizzy, lost control, and the vehicle turned over. Luckily he survived.

Gerald graduated with a BSc in Agriculture in 1947. Honours courses had been suspended except for returning servicemen. In that year there was an opportunity for final-year students to compete for the Steven Scholarship in Agriculture, and Gerald beat the class swot, a girl, to the prize. His award was ninety pounds, a handsome sum of money in those days. He spent most of it visiting animal research stations in Denmark. His friend Ewan Stewart decided to accompany Gerald to Denmark at his own expense. This was much to the surprise of both Gerald and Ewan's family. At times on the trip, Ewan would disappear for a day or two without explaining why. It seemed that Ewan was more successful than Gerald at meeting, or seeking out, attractive girls.

In the genetics class, when the examination results were in, the lecturer made an announcement: 'So, Gerald Wiener, you have come top of the class. You are the only student to gain a distinction. Congratulations!'

Embarrassed, Gerald smiled broadly; he didn't know what to say. This man, Hugh Donald, his genetics lecturer, an overawing figure to some students, was a very elevated figure in the young man's eyes.

'So,' he continued, 'I would like to offer you a post as my assistant at the Institute of Animal Genetics.' Gerald felt flattered and accepted without hesitation. This was a dream come true.

Chapter 9

Meeting His First Wife

During these years, following their meeting at a Saturday night dance in the Students' Union, Gerald had become close friends with the girl who was to become future wife. Sheila was a student at Edinburgh, a highly intelligent girl, who had come close to the top mark of an earlier civil service examination. As a result she was offered a job with the Foreign Office, reserved for only the best candidates, and had the opportunity to make this a lifelong career. Unfortunately, in those days, married women were not allowed to work for the Foreign Office or even to be transferred to another government department. However, as love was in the air, Gerald and Sheila were married. This was a happy occasion, and Gerald was delighted that Ruth Spooner, his 'aunt' and benefactress from his teenage days, travelled up from Oxford to be with them. And of course his mother was there, full of pride at the wedding of her fine, clever, handsome son. The ceremony took place at a Registry Office in 1952, and afterwards there was a small luncheon for the reception party in a hotel in the New Town area of Edinburgh. This was in Charlotte Square, which Gerald came to regard as, architecturally, the most magnificent of squares in Edinburgh. Here the happy couple were surrounded by their small band of family and friends.

The first three years of married life were spent in rented rooms. The first of these, five minutes' walk from Gerald's job, turned out to be a big mistake, amusing in hindsight but not so at the time. Rented accommodation at a price they could afford was hard to come by in Edinburgh. Wanting to be assured of a place to live once married, the two rooms 'with shared use of kitchen' were rented two months ahead of the wedding and the rent duly paid. As soon as they moved in the trouble started. The landlady, Mrs Macdonald, an elderly widow living on the ground floor, would come out of her room to listen for any noise from the Wieners upstairs. Even a radio playing music softly would be cause for a shout to turn it off. Worse was the use of the shared kitchen. If Sheila or Gerald were using the gas cooker and temporarily left the kitchen, leaving the food to cook, Mrs Macdonald would come into the kitchen and turn down the gas or even turn it off completely. Gerald, coming back to the house for a snack lunch (when Sheila would be out at work), would be told gruffly that using the kitchen at lunchtime was not part of what Mrs Macdonald expected. The daughter and son-in-law of their miserable landlady occupied the attic floor of the house, which fortunately had its own small kitchen. Gerald and Sheila became friendly with that couple and they confided that Mrs Macdonald specialised in letting out her rooms to young couples planning to move in at a later date, so that she received rent without having the tenants in the house. Once they had arrived she would make herself unpleasant enough to get them to leave so that she could start over again with another young couple paying rent in advance. For Gerald and Sheila, their departure came after six months.

Their next abode, some miles out of Edinburgh, could not have been more different. The couple, the Macleans, and their grown-up son became the best of friends and were helpful to a fault. But living out of Edinburgh required a car and one at a price they could afford. That came to be a 1927 Austin 7 for the princely sum of thirty-five pounds. The old jalopy had been converted to look like a two-seater racing car (just in looks, not in performance) with an aluminium body – only the chassis, wheels and engine of the original car remained. It seemed a fun buy, but it nearly ended in disaster on the first outing. The seats were just above ground level and approaching cars seemed gigantic to the driver. Driving it home at dusk after the purchase, Gerald felt blinded at eye level by the headlights of an oncoming bus. He swerved into its path, stopping a mere five yards in front of the bus – which, fortunately, had also stopped. An irate driver told Gerald off in no uncertain terms, but Gerald didn't need that telling as he was shaken enough. The car became something of a family pet but after a couple of years was sold for fifteen pounds and replaced by a square black box of a car with four doors and a roof, a luxury that kept the rain out – a 1934 Hillman.

After two more years Gerald and Sheila managed to purchase a top-floor, reasonably modern flat in Edinburgh with a ninety per cent mortgage. The main source of heating was an open fire in the living room and Gerald never ceased to be amazed that the coalmen would lug hundredweight sacks of coal up three flights of stairs (no lift) and deposit the coal in a cupboard in the flat. They were strong men accepting their task with only the occasional moan to extract an extra tip for their trouble

Chapter 10

The Scientist and the Big Sheep Experiment

Gerald's dream of an assistantship with Hugh Donald at the university took an unexpected turn. It so happened that Hugh Donald himself was offered a position to head animal breeding research with the Australian government body CSIRO, responsible for overseeing and funding research. However, he told Gerald not to worry as a new organisation, the National Animal Breeding and Genetics Research Organisation (NABGRO, as it was called at first) was establishing its headquarters in Edinburgh (see Chapter 27) and was sure to give him a job.

As it turned out, Hugh Donald was unable, for family reasons, to take up his position in Australia and was himself appointed to head up the animal breeding research in the newly created organisation, under the director, Professor R.G. White. Gerald recalled wryly that Donald and he were interviewed, separately of course, on the same day for posts in NABGRO, but of course in very different grades – one at the top of the tree, and Gerald near the bottom as the first junior scientist in the grade of Assistant Experimental Officer (on probation).

And so, Gerald and Hugh, the clever, good-looking New Zealander, were thrown together after all at the start of the great animal breeding adventure. Almost the first question Gerald asked of his boss was whether he should change his name from 'Wiener' now that he was starting out on his career. Hugh's typical retort was, 'No, I don't think so, Gerald. A foreign name has never been a drawback in science.'

Gerald did not take up his post until September 1947. It could have been almost three months sooner, but he was earning more money from his summer jobs. His starting pay in the research organisation was £280 per year before tax – rather meagre even then. Those of his contemporaries who had chosen to study at the Institute of Animal Genetics before applying for a post in research were financially slightly better off, as their bursaries were not taxed. Gerald, however, had always had it drummed into his head by his mother that he should earn his living and she did not regard a student bursary as an income.

One of his summer jobs, indirectly related to the research post he was about to take up, was with the Imperial Bureau of Animal Breeding and Genetics, as it was then called. Their staff produced abstracts of the world's scientific publications in animal breeding and genetics for publication in the monthly *Animal Breeding Abstracts*. Gerald had a temporary job as an abstractor. It was a job that he found less than exhilarating, but it helped to pay for his keep and it was related to animal breeding research. His other summer jobs were even more profitable – they were the ones to do with seed-potato growing.

For his own research Gerald was advised by Dr Donald to study the population structure of a major breed of cattle and the implications of such structure for genetic improvement. This was to be desk-based research. No farm facilities had as yet become available for any other kind of experimentation. The study was, however, also chosen as a subject suitable for a PhD thesis. Accordingly Gerald was registered as a part-time PhD student (part time because the work was not done in a university department). Professor Waddington, as head of the genetics department in the university, was his notional supervisor. Gerald was never asked to see him after the time of his enrolment and Hugh Donald was named as the de facto supervisor for his PhD.

The Ayrshire cattle breed was at that time arguably the best milk-producing breed in the country, until usurped by the Holstein-Friesian breed. Gerald's studies involved frequent trips to Ayr and to London to collect information from the pedigree herd books of the Ayrshire breed of cattle and from other publications.

Two connected events occurred to interrupt the smooth flow of Gerald's research and the other duties Donald had imposed on him: helping to set up and later supervise the records section responsible for maintaining all farm and research data that was to accumulate over the years.

The first of these events came out of the blue in 1949. Gerald received an offer from a former student friend (not an agriculture student), Gary Wildman, then in Israel, to take up a job as farm manager of a kibbutz. Gerald didn't think it was really his scene but he did go to see the Director of ABRO, Professor White, and his own immediate boss, Hugh Donald, to tell them of the offer. They immediately asked him to stay, promising to recommend him for promotion to the Scientific Officer grade, which would then be permanent and pensionable, and give him an increase in pay. And so it happened.

Soon thereafter Professor Stephen Watson of the University of Edinburgh School of Agriculture suggested to Gerald that he might like to apply for one of the several lectureships being created in his rapidly expanding department. He added, however, that Gerald would have to suspend, perhaps indefinitely, his research and PhD aspirations as the new posts would take up all his available time. Gerald was unwilling to do this and decided that the challenges of a research career and the excitement of adding to knowledge outweighed, for him, the attractions of teaching. Therefore he did not apply, but several of his friends from his student days, among them Ken Runcie from his own class, did so and made great careers in the university and the College of Agriculture.

Thus, Gerald continued with his research into Ayrshire cattle populations for the first three of his years in ABRO. Submitting the work for a PhD had one disadvantage: according to the rules it prevented publication of the results in scientific journals. Others had become interested in this type of study as a result of coming to chat with him and, unfortunately, one of these studies was published before his. This upset Gerald as, in science, being the first in a field secures the most recognition, but he could do nothing about it. In later years, however, the researcher in question did give Gerald the credit for having been the first to show how the cattle population structure and dynamics affected breed improvement.

A couple of years after Gerald obtained his PhD degree, three eager young scientists, smiling, joking and in those days occasionally smoking, were leaning over a large table covered in papers. This was the outset of a year of planning for an experiment that was to last twenty years. The three scientists were Gerald, who had instigated the plans, Clair Taylor, a brilliant mathematician who was interested in growth in animals, and Alan Dickinson, who had studied crossbreeding in cattle for his PhD. They set about to design a major long-term experiment with sheep.

Dickinson was tall, agile and slightly stooped, as if he had too much to do and couldn't find time to straighten up. He was keen as mustard in his scientific pursuits. In this he was the same as Gerald; like a terrier with its prey, he would not let problems go. Eventually Dickinson concentrated on the disease of sheep known as scrapie. He became an expert and the director of a large research unit into this and similar slow-developing diseases such as BSE, also known as 'mad cow disease'. In his private life, he was a quiet-living man with a lovely family of three girls and two boys. He and his wife were Quakers, and they did good work in the local community. Both were keen gardeners.

Clair Taylor, the brilliant mathematician, was a more sophisticated type. He was quick and sharp in his thinking. The three got on well. They made each other laugh at times and they spent many hours over many weeks discussing and planning the great sheep-breeding project.

Their intention was to be able to answer, for sheep at least, the age-old questions about nature versus nurture, though they didn't phrase it that way. In their more scientific language they wanted to study the relative importance of heredity, maternal environment and other aspects of the environment for a whole range of performance-related characteristics in sheep. The complexity of the design of the experiment needed more than a year in the planning before approval was given for the go-ahead. After all, 'long term' for a breeding experiment meant a plan that would be valid for twenty years. Several breeds of sheep that differed in many of their characteristics were chosen, and Blythbank Farm was to be the location of the work.

The farm had been purchased by the research organisation for the purpose of field studies in breeding. The recently appointed farm manager, Jimmy Harris, and his family were there to meet Gerald when he first drove through the gentle countryside of the Southern Uplands, twenty-odd miles from Edinburgh, and up to the stone farmhouse.

It was a warm June day and tea and sandwiches were prepared. Harris was a kindly, God-fearing man and very good at his job. He and his wife were accustomed to fostering children whose families were in trouble. Slightly apprehensive of the young scientist from the big city, he listened carefully to what the sheep project was to be about.

Gerald explained that it was proposed to have two hundred and sixty ewes, with lambs and young stock over and above that, as well as a lot of rams. Jimmy thought that would be a lot to handle and wondered about the breeds involved, as that would affect management.

It was proposed to have three hill breeds: Scottish Blackface, South Country Cheviot and Welsh Mountain, each quite different in many of their attributes. These were to be carried through the whole of the experiment. But in addition, for the first few years for crossbreeding, there would also be three lowland breeds – the large Lincoln Longwool, the meaty Southdown and the Merino as a wool breed. The large differences in the characteristics of the different breeds were crucial, Gerald explained, to the needs of this experiment.

The farm manager was beginning to worry how these breeds would be used, and what the breeding plan was. There were to be different stages. The hill breeds would be bred pure but also used for crossbreeding. For example, Blackface ewes were to be mated to Welsh rams and vice versa. Thereafter things would become more complicated by introducing close inbreeding.

Jimmy Harris noted these facts and in his usual meticulous way wrote it all down in his notebook. He did not like the idea of inbreeding, as this would depress the output from the flock. Gerald agreed that he and his colleagues were aware of this consequence of inbreeding but it was justified in order to show how the different traits, like reproduction and growth, were inherited.

That left the use of the lowland breeds to be explained. They were there to provide additional variety to the study of inheritance of the performance characteristics of the breeds. For example, was hybrid vigour from crossbreeding important? The Lincoln Longwool, the largest of the British breeds, would also be used in a separate trial. They would be mated to the Welsh Mountain – one of the smallest sheep breeds– and the Welsh Mountain to the Lincoln. In a sense this was harking back to what Gerald had seen and experienced at the research farm in Cambridge years earlier. He recalled how

Hammond's Shire/Shetland cross had demonstrated the importance of maternal effects, distinct from genetic inheritance. But with these two very different sheep breeds and the larger numbers involved it would be possible to quantify the size of the maternal effect rather than only demonstrate its existence.

Jimmy's anxieties mounted when he learned that the experiment was planned to span twenty years, but Gerald reassured him that this was to be the first such experiment of its kind in its objectives and that useful results would emerge at each stage of the breeding work, not just at the end.

Gerald invited Jimmy to let him and his colleagues, Alan and Clair, know of any problems he foresaw and what he would need in terms of extra staff and facilities. This would be incorporated into the final plans before they were finally approved and given the go-ahead.

Some of Gerald's excitement was picked up by the older man, who himself warmed to the idea of a trail-blazing experiment on the farm of which he was manager. Future meetings with Jimmy were arranged and these would include Clair and Alan, who had been closely involved in the design and planning of the experiment. Gerald had come out to the farm only as a first visit to introduce the ideas and for Jimmy to adapt to the proposed plans. Jimmy wanted to know whether the three collaborators all had similar scientific interests. Their interests overlapped but Clair had a very special interest in the complexities of growth and Alan in the effect of the mother's environment on her young. By contrast, Gerald admitted, he was something of a jack-of-all-trades, an 'animal production' man at heart, and therefore interested in all aspects of performance of the sheep. Gerald and Jimmy parted that first auspicious day, having made the beginnings of what was to be an almost lifelong friendship, not just a work relationship.

The sheep in the experiment needed frequent measuring and weighing and so the three scientists made regular trips to Blythbank

Farm, which kept them in close touch with the practical side of the work and with the conscientious experimental officers and shepherds, responsible for tending and recording everything to do with these sheep. The staff at the farm worked long hours, often beyond the call of duty.

One event that had potential implications for the study has to be mentioned here, although it occurred in 1961, a few years after the start of the experiment at Blythbank Farm.

In May of that year, Gerald had a visit from a young Finnish lecturer on animal breeding, Kalle Maijala, an event that was to have far-reaching consequences for sheep breeding in many parts of the world. The conversation took place in Gerald's room in ABRO. Kalle and Gerald got on well together. Kalle wanted to hear about Gerald's studies and for a while they discussed the long-term sheep experiments that had recently started at Blythbank Farm. Then Gerald wanted to know about Kalle's work. Kalle remarked that he didn't want to be boastful, but he had to say that in Finland they had a breed of sheep where the ewes produce more lambs than any breed he knew of in the United Kingdom. Gerald presumed that Kalle was speaking of the birth of many twins, but his friend was insistent that the breed produced many triplets, and that even quadruplets were not unknown.

This was quite startling information to Gerald, immersed as he was in the attributes of different breeds of sheep. When he asked for the name of the breed, Kalle told him that they were called Landrace sheep, and that they were the best sheep in the world.

Gerald thought about it the next day when he was visiting the farm to help with the measuring of the animals. He had taken his dog, Cindy, with him, rather than leave her in an empty house during the day. In the car going back to the Edinburgh he spoke to the dog, which stood panting good-naturedly on the back seat.

'What do you think of that, Cindy? Twins and triplets, sometimes even quads. We should do something about that, shouldn't we?'

When he got back to King's Buildings, he called Alan and Clair. They listened to his idea of acquiring some of these Finish Landrace sheep for ABRO so that they could be trialled alongside British breeds. Because the three were intent on studying the genetic causes of breed differences they immediately wanted to include this new breed in their experiment.

Sadly for them the director refused permission, but, astute man that he was, he saw the potential for the sheep industry. He asked Gerald to write to his Finnish friend for further details, which he did on 8 June 1961. And the rest, as they say, is history – at least in sheep breeding circles. The Finnish Landrace was imported to the UK, initially for use in research at ABRO, led by a young animal physiologist, Roger Land, who was appointed for the purpose. Later, Roger was joined by Ian Wilmut and Bob Webb, each working, albeit collaboratively, on different aspects of the control of reproduction. All three achieved glittering careers.

Subsequent to the importation of the Finnish Landrace breed to ABRO it was imported by many countries, including the UK, to bring some of its genetically high lamb production into other breeds. Kalle Maijala remained immensely proud of Finland's special sheep breed and often referred back to that conversation with Gerald. Gerald, in turn, remained pleased that a chance event had led to lasting consequences. Gerald and Kalle became lifelong friends.

The crossbreeding and inbreeding experiment with sheep continued, against the predictions of the sceptics, for the intended twenty years, to provide unique insights into the consequences of breeding practices. Alan Dickinson had withdrawn from involvement with the sheep experiment after three years and Clair Taylor a few years later. However, Gerald was joined by other assistants and collaborators. By far the most influential of these was John Woolliams, for whom Gerald often expressed his admiration in later years. Gerald recorded an interesting story about the few weeks between John accepting the post and arriving in Edinburgh. John had just completed a degree in Cambridge in mathematics and statistics, an important requirement for the job, but he had no knowledge of genetics. He was personable and very anxious to learn. With the help of a textbook he absorbed in a matter of weeks what took others a year or more – that is how clever he was. He became recognised in later years as a leading professor of animal breeding and genetics.

Gerald continued to work hard on other aspects of the organisation's research programme. The studies of the Ayrshire cattle population had occupied most of his time from 1947 to 1950. His involvement with the records section, employing at its height a staff of seven women to deal with the mass of records from cattle, sheep, pigs and also, for a time, mice, changed, and eventually ended, with the purchase by ABRO of a computer the size of a room. It did in fact require an air-conditioned room for itself. By today's standards it was a model out of the ark but it was a great advance on the earlier use of punched cards and desk calculators. The computer itself needed to be looked after by specialist staff, and the more advanced methods of analysing the data from the large number of different experiments required dedicated statisticians, not Gerald.

From about 1949 onwards Gerald began to be involved with aspects of several of the cattle experiments that had been started under initiatives from Hugh Donald. The most important of these was to look, in different farm situations, at the complex issue of nature versus nurture. In this particular case it involved using identical and non-identical twins as well as pairs of half sisters to unravel the story. By way of explanation, these pairs of animals share different proportions of their inheritance: for example identical twins share all of it, half sisters only a quarter. So, differences in performance between members of a pair, say in growth or milk production, can be related to the differences in their inheritance. On a more limited scale, identical twins had for long been regarded as useful in studies of intelligence in man – and that no doubt helped to foster the ideas for the cattle studies.

In those days, the decision on whether twins were identical or not depended on the visual comparison of a large number of individual traits. Gerald developed a useful addition to this battery of assessments. He argued that as identical twins in humans had very similar fingerprints, so cattle twins might have similar muzzle patterns. He developed a way of comparing patterns by taking plaster casts from the muzzles and then analysing and comparing them. There was a much closer resemblance in the patterns of twins thought to be identical than those of non-identical twins or of unrelated pairs of calves. This was not an absolute route to determining whether cattle twins were identical or not, but it did help significantly in the diagnosis – and the method of obtaining the pattern from the animals caused some amusement.

There was also a lighter side to the lives of young scientists. Encouraged by Hugh Donald, Gerald and most of his colleagues became members of the British Society of Animal Production. This took him away from home for annual scientific conferences and in some years to other events organised by the Society.

In 1953 Gerald had the first taste of one of the privileges of the life scientific by being allowed to attend the International Congress of Genetics held in Bellagio on the beautiful Lake Como in Italy. On that first occasion he was just an attendee and not a contributor to the proceedings. His wife was able to join him for a short holiday in Florence after the conference had ended. More exciting travel was to follow.

Chapter 11

To the USA

The year was 1956. By then Gerald had been promoted to Senior Scientific Officer with an increase in status and pay. In that year he was awarded a Kellogg Foundation fellowship to spend six months (though it ended up being eight) for post-doctoral study and travel in the USA. At that time the two largest centres with the best-known animal geneticists were in Ames, Iowa and Madison, Wisconsin. Charlie Smith, one of the scientists on ABRO's staff, had been a student in Ames and John King (another of ABRO's scientific staff) had been in Ames a year earlier, also on a Kellogg fellowship. Hugh Donald, by then ABRO's director, thought it best for Gerald to go to Madison in order to broaden ABRO's joint experience of American thinking and teaching in animal breeding and genetics. Gerald obviously concurred with this plan, hoping also that the approach to animal breeding in Madison might be a little less mathematical than it was in Ames. In that regard he was to be disappointed, though the methodology was very different, as Hugh Donald had predicted.

Gerald set off for the USA in February 1957 on the Cunard liner RMS *Saxonia* – a luxurious ship able with a capacity of 1,000 passengers and a crew of 600. The first night out, travelling from Liverpool to Cork, was magical for the young man. As a Senior Scientific Officer, he was allowed to travel cabin class, shared with a young army officer. Never before had he seen a dining hall with so much luxurious food. With a great sigh of expectation, Gerald eyed the choice of meats and cheeses, the beautiful desserts and the steaming, fresh coffee.

Things changed after the ship left Cork for New York and a violent Atlantic storm battered the ship. Quickly the happy atmosphere changed with the banging and clashing of furniture and movable objects flying across the decks. The dining hall was soon an unbelievable shambles as the storm gained in strength. Most passengers on the ship were seasick for a couple of days, but Gerald continued to be sick for several days longer. Completely miserable, he stayed in his cabin, with frequent doses of medication from the ship's doctor. Because of the bad Atlantic weather the *Saxonia* had to sail further north than usual. One day the juddering from the ship's engines subsided and the doctor urged his patient to come on deck, assuring him that he would feel much better if he left his cabin, as the weather was much improved.

He found it difficult to leave his sickbed, but the effort was worth it and proved to be very good advice. The day was beautiful. Bright sunshine, blue sky, calm sea, the ship floating among arctic ice floes as far as the eye could see. After a few hours the ice drifted away and the ship continued on to make an unscheduled stop in Halifax, Nova Scotia (Canada). Passengers were allowed on shore for a couple of hours and they were happy to have firm ground under their feet.

From Halifax the ship sailed on to New York, arriving there five days later than scheduled. Gerald's wife, Sheila, had flown out from home to meet him in New York and welcomed him at the quayside, worried by the delay in his arrival. A nice Jewish man, Mr Zoegall, whom she had met on the flight over from Scotland had come to her rescue, finding her a hotel room and generally looking after her. He also was at the quayside and although they never met again they corresponded and remained friends for some years until he died.

From New York Gerald and Sheila took an overnight train to Chicago, arriving there early in the morning. They requested rolls and coffee for breakfast at the station counter. This resulted in coffee and very sticky, sweet buns. It was the first but not the last time they learned that Americans use different words for common objects.

On arrival in Madison, Wisconsin, they were met at the station by Professor Chapman, who was Gerald's appointed mentor. He took Gerald and Sheila immediately to his lovely home in a leafy part of Madison and introduced them to his charming wife and two daughters. They sat down to what was, for the two newcomers, a very unusual lunch. Besides the usual meats, cheeses and salads, there were also breakfast cereals, jam and all the things they would have associated with breakfast. Professor Chapman, or Chappie as he was always affectionately called, proceeded to tell them that this was the informal way that they had in the USA and then also went on to give them a mini tutorial on the many words that meant something different in the USA. Chappie had been born in England but was, in all his ways, as American as could be.

When Professor Chapman declared that they must stay with him for a few days until accommodation could be found, Sheila and Gerald protested that he was being too kind, and that they did not want to inconvenience him. But the good man would have none of it. He insisted that he had been looking forward to their arrival, and that tomorrow he would take them to a few addresses to enquire about a place for them to live during Gerald's time at the university. Then he and his wife would take them round Madison and show them some of the sights. They were delighted at this suggestion. Gerald said that they had seen a beautiful lake as they drove into the town. Chappie replied that they had seen only Lake Mendota and that there were another three lakes around Madison. He would show them these others the next day and promised them a walk down State Street in the centre of the town. The two new arrivals to the country were happy at the prospect of seeing the city with such kind and welcoming people.

The next day, they walked around the environs of the stunning Wisconsin State Capitol building. As they scanned the enormous white building, Chappie gave a running commentary on the many different kinds of stone used in the building and its other features. He was enthusiastic about his city and proudly pointed out that at the top of the building, the statue's right hand is raised in salute, proclaiming the state motto, 'Forward'. And in her left hand she holds a globe with an eagle on top.

As they walked around, their proud guide told them that this Capitol building had the largest granite dome in the world. The city did not allow the building of anything taller than that within one mile. And another thing to boast about was that forty-eight per cent of Madison adults had at least a bachelor's degree from university. Furthermore, there was the greatest number of PhD graduates of any comparable town in America. This was but a foretaste of the enthusiasm and zest that Chappie brought into all things, including his career as an animal geneticist.

They had a cheery lunch in a café on State Street, and soon were shown round a house which was two or three miles from the city centre. They settled on this rather higgledy-piggledy place, the owner a slightly eccentric woman of about thirty-five. She was known as Francie, and she was a farmer's daughter. She left them to get on with their own affairs and showed no wish to be friendly or sociable with them. Possibly their British accents and manners put her off. She left the house early each morning to go to work, which was definitely on a farm. Each evening Sheila and Gerald had to hide their laughter at her appearance – covered in mud and smelling to high heaven of cow dung. Sometimes she brought home her boyfriend who stayed the night with her, but she did not discuss this with them.

Five days a week, Gerald travelled two miles by bus to the lovely campus of the university, beside Lake Mendota. The bus was usually almost empty as most people took their large cars. He then made for the relatively small genetics building. Here were to be found a host of world-famous geneticists. This building was squashed between the large agricultural engineering and biochemistry buildings, as if to make use of every inch of space. Not far away was the mighty and imposing Agriculture Hall where Gerald often went for seminars and discussions. He attended many lectures and tutorials, and found that their ways of looking at data and their ways of analysis were very different from those of Edinburgh. This, of course, was why he had been sent there.

A few weeks later, after Gerald and Sheila had settled in Madison, a phone call came to them from an old University of Edinburgh friend. This was Jim Henderson who, unknown to them, was in Madison on a post-grad fellowship in crop production. It turned out that Jim and his wife, Pat, had come to America with their two young sons, and that they were unhappy with their accommodation. The Wieners' landlady, Francie, was persuaded to rent out the whole of her house to the two families from Scotland.

This new living situation was helpful for sharing housework and cooking, as well as for the pleasure of having friends in the house, and no smelly Francie returning in the evenings. However, it was also fortuitously helpful for Gerald, as a problem had arisen with Sheila's job. She had found a position as the personal assistant to the Secretary of the Civil Service Union, headquartered in Madison but due to transfer to Washington, DC. Gerald suspected that there might be a mutual attraction between Sheila and her boss, who was trying hard to persuade her to move to Washington, in spite of the fact that she might remain there for only a couple of months. Sheila was keen to go. Gerald fell silent at this prospect for some time. Eventually, with the help of Pat and Jim, he managed to persuade Sheila to stay in Madison until August, when his studies would be finished, and they would take a trip west to San Francisco.

The genetics department at Madison had achieved some fame through pioneer geneticist Sewall Wright. Wright, from Madison, and R. A. Fisher and J. B. S. Haldane, from Cambridge in England, were regarded as the founders of the mathematical and statistical elements of animal population studies. They were followed by other 'giants' in the field of animal genetics, principally Professor Lush at Ames, Iowa and, though slightly less well known, Chapman at Wisconsin. Lush as well as Chapman had interpreted earlier theoretical studies and extended them to actual animal breeding plans. Lush, however, had become internationally regarded as the doyen of animal genetics applied to breeding through his book *Animal Breeding Plans*, a standard textbook for many years. He had beaten Chapman to the draw, so to speak, as Chapman's own 'animal breeding plans' remained confined to his large volume of lecture notes for students.

During Gerald's time in the genetics department in Madison he attended many lectures and tutorial groups with other post-grad and post-doc students. Many of the ways of looking at data and the forms of analysis were new to him, and were indeed different from those adopted for the most part in Iowa and Edinburgh. He was used to the more relaxed attitude in ABRO where the scientific staff would often sit for a long time after the formal morning coffee break to discuss animal genetics and work-related issues – often in the company of Hugh Donald, by then the director. Such informal intrusions on the day's work were alien to the way they operated in Madison. The prolonging of coffee time discussions beyond the allotted fifteen minutes, often initiated by Gerald, was regarded initially as a waste of time by Prof. Chapman. But a longer time spent over coffee with profitable discussions became tolerated, perhaps even liked, and after a short time no longer regarded as 'un-American'.

Wisconsin, being the Dairy State of the USA, had a large university dairy department where they made and sold wonderful ice cream. Even years later Gerald felt he could still taste it. The university genetics department also had a sheep section (with Art Pope at its head). Sheep were not a common farm animal in Wisconsin and Art Pope and his department made it their mission to promote sheep production. Gerald recalled accompanying Pope and others to sheep 'open days' in the north of Wisconsin. To engender interest in the meat they made barbecued lamb burgers. To someone from Scotland, this did not seem the best way of advertising how good lamb meat could taste if roasted. Gerald had to give a talk to farmers and others at a meeting during those few days, but what the farmers thought of hill breeds in Scotland he never discovered.

Towards the end of his six months in Madison, Gerald purchased an old Pontiac car that had been traded in by Chappie, who told him that he would not sell it to him as he could not vouch for its roadworthiness, adding that it was a long drive to San Francisco. But the car dealer offered it to Gerald 'unwashed' for ninety dollars. Gerald jumped at the deal and he and Sheila looked forward to using the car for their journey across country to the West Coast.

Around the beginning of March, Gerald became friendly with a librarian who helped find material for his studies in the science library. He was Ned O'Neil, and he and his wife, Anne, were from San Francisco. They were in Madison for a few months for a course of study that she was doing. One Sunday, when the Scottish couple were walking by the lake, intending to have a picnic, they came upon Ned and Anne in a boat close to the shore. Dark-haired, handsome Ned stood upright in the boat, smiling and making a comical attempt at singing 'O Sole Mio' in a loud falsetto voice to the reclining Anne. They stopped in amusement at this picturesque performance on the blue, slightly choppy waters of Lake Mendota. They were invited to join the couple for a picnic and met them at the marina. On that beautiful, sunny spring day, the four shared a picnic. This was to be the first of many days around the lake and on the campus that they spent together.

Sad to say goodbye to the fine friends that they had made in Madison, to the beautiful city, and the wonderful lakes and woods that surrounded the campus, Gerald and Sheila set off in August on what was for them an epic journey. En route they visited several renowned animal breeding and animal production departments: Ames, Iowa; Fort Collins, Colorado; Laramie, Wyoming; Corvallis, Oregon; Davis and Berkeley, California; and the famed Beltsville experimental station in Maryland on the way back. In several of these institutions Gerald renewed his acquaintance with men and women whom he had met in Edinburgh when they had visited ABRO and the Institute of Animal Genetics. As this trip took place during the university vacation period he was spared from giving formal talks, as Chappie had warned. Instead he just had informal chats about his work – usually during a coffee break.

Gerald and Sheila travelled with a pup tent lent to them by their friends Ned and Anne O'Neil. From time to time they camped overnight to save money. One such night, the tent was pitched in a small park not far from a railway line. That is when they discovered that American freight trains were enormously long.

The journey westward took them past the Great Plains, towards Colorado and Fort Collins on the edge of the Rocky Mountains. The car that Gerald had grown to love – so much larger than his first car, the 1927 Austin 7 – had developed a bad engine fault on the way to Fort Collins. With the help of two old friends, Professor Stonacker and a former Edinburgh graduate, Tom Sutherland, now lecturer at the university in Fort Collins, a garage was persuaded to repair the engine, rather than trade in the car, the norm for most Americans. The lovely Pontiac was now to be restricted to a top speed of fifty miles an hour.

The drive to Laramie in 'the wild west of Wyoming' brought them to another old friend. Larry Parker and his family had spent six months in Edinburgh and had visited the couple in their home there. Larry was a wool expert and had spent much of his time in ABRO with the two people there who specialised in wool. Gerald and Sheila spent an overnight with them and, to Gerald's great embarrassment and the Parkers' dismay, he drove off on the wrong side of the road whilst waving goodbye. The Parkers' waving was more frantic – but fortunately the dirt road was deserted. They then crossed the Rockies and made their way northward into Washington state (Sheila had friends in Spokane from her school days in Scotland) and then down the West Coast through Oregon with its giant redwoods into California. Walking through the trunk of a giant Sequoia that had been hollowed into a gateway was quite memorable. The lovely Pontiac then lasted them into California. Two more important university visits followed – to Professor Lerner, a renowned geneticist and friend of Hugh Donald in Berkeley and Professor Bradford in Davis who had close ties with ABRO. This was the kind of engagement with the United States desired by the Kellogg Foundation. It made good material for the report that had to be submitted to them.

But there was a sad end for the Pontiac. California law prevented cars from out of state being sold for some months after arrival, unless sold as un-roadworthy for scrap. So that indignity befell the car Gerald had grown to enjoy.

Chapter 12

Finding a New Family

Gerald's biggest hope in San Francisco was to find his father and any family he might have. The message he had received via the Red Cross some eight years earlier had made it likely that this, if anywhere, was the place to look. He had always regretted taking his mother's advice not to accept his father's invitation to make contact.

San Francisco was an exciting place to explore in the 1950s. The two travellers spent some days taking in the usual tourist sites in this bustling, crowded city, including the not-to-be missed ride on a tramcar from the uphill terminus down to the harbour. Endless hours could be spent on the streets of the city, viewing the unique buildings, exploring areas such as Chinatown, and taking in the people of many nationalities who had made the city their home. But soon Gerald could no longer restrain himself from looking for the father he knew had emigrated to this city.

Leaving Sheila to do some shopping, he got hold of a phone book and started looking up the addresses of anyone called Wiener. Finding quite a number with that name, he called in person at the first two addresses he found – only to have the doors slammed in his face the moment he had said 'excuse me'. Perhaps they thought he was a beggar – or worse.

He then took to phoning instead. The second such call was answered by a woman who, when Gerald introduced himself, exclaimed a big 'Mein Gott'. It was Ursel, Paul's wife. When she recovered from her shock, Ursel found herself able, in her strong German accent, to invite them to her apartment, and gave him the address.

Nervously Gerald and Sheila approached the door of the apartment. The bell was answered by a tall, smiling man, who looked just a little older than they were. They stepped inside to find several people gathered together, and all eyes were turned on them. The oldest lady there was Gerald's aunt Thea. She came forward, introducing herself. It was obvious that she was as nervous as he was. She hugged first Gerald and then Sheila. In shaky voice and accented English, she told him that she remembered him as a small child, about two or three years old. Seemingly he was very cute then, and he used to babble out sentences while he was running around. She said that he spoke so fast that he got the words all mixed up. Thea turned to one of the other women in the room, slightly younger than herself but very attractive, and said, 'See, Ursel, he has the look of Paul, don't you agree?'

Ursel was the one who had answered Gerald's phone call and from that he knew already that she was his stepmother. Ursel agreed about the looks and tears came into her eyes as she too hugged the visitors. The others in the room were Thea's daughter, Vera, and her husband, Hans, who had answered the door. A fair-haired, handsome boy of perhaps ten years old sat quiet and glum. He was Ursel's son Jerry – Gerald's half-brother. Ursel told that she had an older boy, Pete, also a half-brother to Gerald, but Pete could not be there that day.

Vera brought in coffee and cakes and when they were all seated, it fell to Aunt Thea to tell Gerald that, sadly, he was too late to meet his father. Paul Wiener had passed away two years previously, of a heart attack.

Gerald's face fell. The broad smile was gone, and disappointment hit him. Feelings were developing inside him at such a great rate.

Such shattering information was hard for him to deal with, especially after the love with which the family had greeted his arrival. Life up until now had been friendly – you got by, you coped. His mother, now in England, was paranoid about his welfare, full of warnings and instructions to study, to work hard, to be careful of women, to watch his spending. He knew his mother loved him, but this well of feeling he was experiencing in this room in San Francisco was totally overwhelming.

His stepmother put her arms around him. 'What a shame! What a great, great, terrible shame! Your father spoke of you often. I know he loved you.' She was sobbing. The raw emotion of mourning had not yet left her. She started speaking in German to her sister-in-law, but he could not catch the meaning. 'I know Paul would have loved to see you.' Ursel shook her head in sorrow, her words emerging sad and soft.

Gerald's aunt told him that his father had written to him as he wanted to give him a home in America. In a croaky voice Gerald explained that his mother made him write back to the Red Cross saying he wanted nothing to do with his father. He tried to explain how his mother was strong-willed and controlled him, even though he had known in himself that he wanted to meet his father. Thea's answer was that Luise must have still been so bitter about the past. Thea and Luise had remained friends, at least in thought, despite Luise's divorce from Thea's much-loved older brother, Paul.

Soon the day reached its peak and words were becoming fewer. Now the young couple were getting restive, as the next morning they would have to catch a plane to New York on the first leg of their return to Scotland. They did not want to leave and nor did Thea want to see them go. Too many memories were flooding back to her mind. When he was a toddler, Thea told him, he had made them laugh and they loved him. Having to say goodbye so quickly upset her. She made them promise to come back to see her again. She said that there was so much more to tell and that he was now part of the family again. Gerald felt that too and promised to keep in touch and some day to return. To have a family at last! This was a day to look back on, a very happy day in Gerald's life.

Following their return from the USA and the pleasure and stimulation of that trip, Edinburgh might have seemed a dull place. But, in addition to its many existing attractions, the city was becoming ever more cosmopolitan in character. Soon after the end of the war was the successful launch and then resounding success of the Edinburgh International Festival of Music and Drama, and the ever-expanding number of Festival Fringe events brought a new life and atmosphere to the town.

For the period of their absence in the USA they had let out their flat in Falcon Road. A few months after their return they bought, with the help of a ninety per cent mortgage, a fine terraced house, built in 1890, in Dalhousie Terrace in the Morningside district of Edinburgh – a nice area of the city but often the butt of jokes about the alleged refinement of its residents, though like most jokes this had only a modicum of truth attached to it.

By this time Gerald's mother, Luise, lived nearer to hand, so Gerald had an early opportunity to tell his mother of his new family in San Francisco. She listened intently, even excitedly, to his story of the meeting with Thea and the others, and of the death of his father. Her earlier antipathy seemed to have disappeared, and she was sorry that Paul had died so young, only in his fifties. Mostly she welcomed news of Thea, whom she had always regarded as a friend. Luise also said that she was glad Gerald now had two brothers, as she referred to Pete and Jerry, something she might have hoped to give him herself, had things turned out differently.

For Gerald and Sheila it was back to work and a couple of years later they welcomed their firstborn, Andrew. Their daughter Ruth was born four years later. Unfortunately, the birth of their children exacerbated Sheila's lifelong predisposition to depression and caused distress not only for her but for Gerald also. For him this meant a need for increased time with the children and for extra domestic chores, although his mother-in-law came to help during Sheila's stays in hospital. Though interrupted by these spells for recovery, Sheila continued in her job, located in Moray House, as secretary to Professor Pilley, Professor of Education at the University of Edinburgh. It did not have quite the glamour of a career in the Foreign Office with the prospect of postings abroad, but she found it interesting work and enjoyed the contact with students.

Gerald for his part became immersed in his research and most particularly in the long-term sheep-breeding experiment, which was underway at Blythbank Farm. He was also invited to take on the role as editor of a scientific publication. It was long recognised that this was a voluntary task of service to the academic community. It was a new responsibility that would eat into the time he could spend on research.

Gerald had been involved for several years as editor for the Agricultural Former Students Association in Edinburgh. That experience got him noticed and so he was invited to become the editor of the Proceedings of the British Society of Animal Production (BSAP), a vigorous society with a large membership including university academics and research scientists, many from industry, agricultural advisors, and even a small contingent of farmers. It became almost immediately obvious to Gerald that here were the makings of a scientific journal devoted to research in all aspects of animal production – something missing from the array of scientific journals in the UK. The scientific content of the society's annual spring conference would provide a good start. Gerald had the support of John Maule, Director of the Commonwealth Bureau of Animal Breeding and Genetics, who was Secretary of the BSAP. Gerald and Maule went, in 1958, to see Douglas Grant, one of the directors of the prestigious publishers Oliver & Boyd in Edinburgh. Douglas agreed to his company publishing a new scientific journal initially named *Animal Production* (later *Animal Science*) with Gerald as editor. The journal was launched in 1959. Over the years it became the leading animal science journal in the UK, with papers submitted for publication from all over the world. The society benefited too, as the journal became a money-spinner for them and thus helped in an expansion of its activities. Specialists in different disciplines joined Gerald as editors after the first few years, initially Professor James Greenhalgh, a renowned animal nutritionist from Aberdeen University, followed in later years by many other distinguished academics. During the twenty-five years Gerald spent as editor, the downside was the large amount of his time it took during both official working hours and at home at weekends. The upside was that his name became widely known as much for this activity as for his own research. The society honoured Gerald by electing him as one of their small number of honorary members.

As a postscript to this, changes in reading and publishing habits and the expansion of online publication led, in more recent years, long after Gerald's retirement, to the merging of *Animal Science* with its European counterparts. It continues with the title *Animal*, in a gesture to please both the French and the British.

There was one other event of note for Gerald at that time which greatly affected him. In the early 1960s, scientists from the renowned Animal Research Institute at Mariensee in Germany had visited ABRO and a return visit was planned. Partly because he was still able to speak German, albeit with a vocabulary limited to that of a twelve-year-old schoolboy, Gerald, along with Clair Taylor, was asked to represent ABRO on this visit. What imprinted that trip on Gerald's memory was not the exchange of research findings and ideas, but the fact that it changed, to a large degree, his feelings about Germans and Germany. From the time he had left the country as a refugee his attitude had been one of almost total hostility. Now he met a generation of young men and women who had not been part of Hitler's gang, whose attitudes to the world were similar to his and who, like him, abhorred the Nazi years and the acquiescence of much of the population in the barbarities of those times. And of course there was the shared excitement in research.

Apart from the trip to Mariensee, there were other conferences and meetings to attend in the years after his return from the USA. He also took an increasing role in the life of the Congregational church at 'Holy Corner' in Edinburgh. And then his scientific career took an unexpected and exciting turn.

Chapter 13

The Lucky Break

In science, breakthroughs sometimes happen quite by chance, maybe through an accident or unforeseen event. Almost everyone has heard of Alexander Fleming's accidental discovery of penicillin. Gerald's discovery in 1964 wasn't in the same league of importance but it also came about by accident and it gave him instant recognition and fifteen years of fame in the agricultural world.

One day in the late spring of 1964, with the sheep experiment going along well, Elfed Hughes, the most senior of the experimental staff at the farm, took Gerald aside to tell him of some problems with the lambs that year. A lot of them had developed swayback and those that did not die had to be put out of their misery. It was hard for the staff to have to do that. Gerald wondered if Elfed had noticed anything unusual about these swayback occurrences, but was unprepared for the answer that the Blackface lambs seemed to have come off worst.

Later, in his room at the ABRO building on the university's science campus, after analysing the lambing records for that year, the slow dawning of what he was uncovering came to him. An excited Gerald called on Clair Taylor, his colleague and neighbour in the room next to him, to tell him that there were huge differences in the proportions affected by swayback between the three breeds and their crosses! Almost half of the Blackfaces had been affected, but none of the Welsh, while the Cheviots had an incidence somewhere in between. Most importantly, Gerald had noted that the incidence of swayback in the crosses between the breeds was close to halfway between those in their parent breeds. It was the best possible evidence that heredity was involved in the occurrence of this disease.

Clair was surprised, his eyebrows raised in puzzlement. Dickinson, hearing the excitement in Gerald's voice from his room on the other side, had joined them. Gerald could hardly contain his excitement that swayback might be inherited, at least in part. His two colleagues, after looking at Gerald's figures, agreed that this was important enough to go and see Dr Donald for permission to investigate this matter further.

But Dr Donald was dismissive of these statistics. He claimed to have seen things like that before and without a trace of sarcasm told Gerald not to let it change his life. A strange injunction as it turned out – because it did. Donald, persuaded by Gerald, allowed him to continue with the investigation.

By way of explanation, swayback is associated with copper deficiency. The lambs at birth, or soon after, have little control over their hind legs and sway when they try to walk, hence the name 'swayback'. As Elfed told Gerald, the affected lambs rarely survive. In areas where the condition is common, farmers are advised to administer copper to pregnant sheep to prevent the condition. But as it does not occur routinely every year, the treatment, apart from the cost, is not always appropriate. Furthermore, copper is an essential trace element in several of the body's proteins in both animals and in man. The implications of a possible genetic component to copper deficiency therefore go well beyond sheep.

They now needed a biochemist to delve deeper into the subject and that came through the strong support of Dr Stamp, Director of the Moredun Research Institute. He encouraged Alex Field, the head of the biochemistry department there, to take up the challenge, and from that point on a major project was born.

Much work was done in analysing the figures of the past years, and Gerald and Alex were asked to present their initial results at the first post-war International Congress on Trace Elements in Man and Animals held in Aberdeen in 1969. Gerald was both surprised and delighted to find himself invited to speak at the opening session, reserved by the organisers for what they regarded as the most novel advances in knowledge since the last pre-war congress. Despite his nerves, his presentation was well received. The chairman of congress, Professor Underwood, a distinguished Australian and the doyen of the subject, had his concluding lecture to the congress recorded in the published Proceedings. They read as follows:

> It is always invidious to select particular papers for special commendation but I don't think anyone will quarrel with me if I mention at this point the very revealing paper by Drs Wiener and Field on genetic variation in copper metabolism in sheep. We have long accepted species differences in nutritional studies. As someone once said 'all men are not guinea-pigs and only a few are rats'. What we have now to appreciate, nutritionally speaking, is that all sheep are not sheep, or at least not sufficiently so to allow confident extrapolation from one genetic group to another.

The implications of the new insight must have had special relevance for an Australian as large swathes of his continent have trace element problems for livestock, sheep especially. This new knowledge allowed a review of both breeding and feeding practices.

The research by Drs Wiener and Field and other collaborators progressed apace with additional research facilities, equipment and staff. For the next fifteen years Gerald, and often also Alex, was invited to further international congresses, taking him to the USA, Germany, Australia and back to Aberdeen in Scotland. On each occasion there were new results to report, finally found to be the influence of genetics on the ability of the sheep to absorb copper from their diet. In practical terms, one breed might die of copper poisoning on the same diet from which another breed would show signs of copper deficiency. The medical fraternity were just as interested in this work, as were veterinarians and livestock specialists.

The work led to around sixty scientific publications with Gerald as either the senior author or as co-author – out of his total of about 120 science publications. In addition he has two books to his name and chapters in others. The genetics of copper metabolism – and hints of the possible influence of heredity with other trace elements and minerals – has become accepted as part of the established wisdom in this field of animal nutrition. But as happens so often in the forward march of scientific knowledge, those who start down a road of discovery are supplanted by a new generation of scientists, some of whom are unfamiliar with the more distant past. In part this may be due to an absence from the internet of many of the older published research findings. Not all researchers take the time or have the means to delve deeper into the archive. Gerald's work is no exception to this rule of a forgotten past, which, nonetheless, is not erased by being buried. He feels that it is reward enough that all breeds of sheep are no longer regarded as equal in their requirements for copper – as was already acknowledged at that congress in 1969, soon after the start of his 'lucky break'.

In recognition of his work, Gerald was elected a Fellow of the Royal Society of Edinburgh in 1970 and a Fellow of the Institute of Biology, now the Royal Society of Biology, in 1976. In the same year Gerald was also awarded the degree of Doctor of Science (DSc) by the University of Edinburgh, his work on copper playing a significant role in this. It was his third appearance for graduation in the McEwan Hall. Gerald recollects with amusement that on this occasion he was the first in the crowd of graduands to be called and was firmly instructed that 'if you get the procedure right all will be fine, but if you get it wrong there will be chaos'. He got it right.

The 'lucky break' also led to another milestone in Gerald's career. Dr John Stamp, the director of the Moredun Research Institute who had so strongly supported Gerald in getting the work on copper off the ground, was also the chairman of the sheep committee of a joint consultative organisation that had been set up by the government. This organisation was to look at research and development in all aspects of farming right across the UK and set priorities for future investigations. John Stamp invited Gerald to become the technical secretary of his committee. As such, Gerald was responsible for writing the committee's reports for six years, and then continued as a member for a further six.

Chapter 14

Ethiopia

Early in 1972 Gerald received a summons to see the director of ABRO. He never knew quite what to expect from Hugh Donald. Usually these meetings would be about some scientific paper he had sent to the director, who had to approve every paper before it could be submitted to a journal for publication. The first paper he had ever given to him for approval, soon after he had completed his PhD, had been torn to ribbons by Donald, who called it 'a dog's breakfast'. But such chastisement had become infrequent as Gerald became more skilled in the requirements for scientific writing. His own experience as an editor helped. There had been one occasion when Hugh Donald insisted on some changes in a paper written by Gerald. When the comments came back from the journal editor, the section of the paper that had been altered to take account of Donald's views was criticised (the reviewer used by the journal editor was of course unaware that the section in question had been recast at Donald's bidding). When told of this episode Hugh Donald said magnanimously, 'Well, I can make mistakes too.' Donald did not, however, repent his strictness with the approval of papers. He had told all his staff, on more than one occasion, that he did not want any scientific papers from his organisation to be rejected by journals or to receive rude comments from referees.

When Gerald knocked on the director's door he was nervous of what might transpire. He need not have worried, as the call was for something quite unexpected. Donald informed Gerald that he had received a letter from Ian Mason, then working as the Animal Production Officer at FAO in Rome (the Food and Agriculture Organisation of the United Nations). Mason had asked if someone from ABRO could be sent to Ethiopia to advise on cattle and sheep breeding. Mason had mentioned Gerald as a possible candidate and Donald wished to know if he was interested.

Gerald's first thought was that he had not done anything like that before and wanted more information on what might be involved. Of course he knew Mason quite well from their time together in Edinburgh and that was a bonus. According to Donald, Mason thought that Gerald had the right research interests and was good at report writing, a big consideration for the bureaucrats in Rome – not that Mason was one of those. Donald assumed that the assignment would be to look at the current state of cattle and sheep breeding and, if necessary, suggest and design a breeding programme.

Donald himself thought that Gerald was a suitable candidate as he was one of the few in ABRO who had worked with both cattle and sheep and he was also good at dotting the i's and crossing the t's. This, coming from the biggest dotter of i's and crosses of t's in the business, could only be a compliment.

Donald had to seek agreement to release Gerald from ABRO for the likely six or seven weeks of the consultancy. The permission was granted somewhat grudgingly as the skinflints at the Research Council decided to withhold Gerald's normal salary. FAO were going to pay fifty pounds per day plus a daily living allowance. So Gerald, in accepting this trip as both challenge and adventure, was not going to get rich on the proceeds (a common assumption about consultancies).

A few weeks later, Gerald, now pumped full of vaccinations, a stock of malaria pills in his luggage, set off on the first leg of the assignment for his briefing at the FAO headquarters in Rome.

On the flight over he happened to find himself sitting next to Bob Orskov – someone he knew well from the annual conferences of the British Society of Animal Production. Bob was an animal nutritionist at the Rowett Research Institute in Aberdeen and a regular consultant in African and Asian countries. He tirelessly spread his message about proper nutrition for livestock. He too would be stopping off in Rome.

Bob wanted to know what he was to do in Ethiopia and by this time Gerald had more information. It seemed that Haile Selassie, the emperor, had personally ordered the importation of some two hundred Friesian heifers and a few young bulls from Holland and Britain. They were bought and airfreighted in at great expense. The majority did not survive many months, succumbing to diseases to which they had no immunity. Staff at the FAO were quite annoyed, as they had not been asked for advice beforehand – even though they were already providing assistance to Ethiopia. But after the debacle following the importation of cattle, the Ethiopian authorities had finally requested help and Gerald was to give it.

Bob thought it would not be an easy job. From his own experience in Ethiopia he had found the people at their ministry rather proud; they liked to think they knew best. Diplomacy was urged, but that was something Gerald had anticipated.

Bob also gave advice about Ethiopian girls. They were said to be very alluring and liked to find a 'sugar daddy' to live with – but mostly those on year-long missions. Gerald was likely to be safe on that score but he noted the warning.

Gerald arrived in Rome for his briefing and the start of what was for him an adventure. FAO consultancies always start and end in Rome. This was the first of many times he would come to this grand city.

It seemed almost like a holiday among the historical and beautiful surroundings. The FAO headquarters are but a short walk from the

Colosseum and not far from the Forum Romanum. Eating late in the cool of a summer evening in one or other of the magic squares of Rome, surrounded by majestic old buildings elegantly floodlit, was a treat. Why, Gerald thought, had he not done this before? Of course the days were filled with meetings, briefings and reading up past studies from Ethiopia. But it was worth it.

It was a beautiful July day with bright blue sky, pleasantly warm, when he arrived in Addis Ababa. He was taken to the Hotel Ethiopia, a large, modern hotel, air-conditioned and with an attentive staff, especially the young women. Gerald felt excited and elated. He was yet to meet his Ethiopian counterparts but the first day in Addis Ababa was to be relished. Always ready for sightseeing, Gerald had a scout around some shops and market stalls. He quickly learned not to run – he had forgotten that he was at 2,300 metres altitude and that it takes time to acclimatise. (He wondered whether, when he got back home, he would he be able to run faster because of all the extra red blood cells in his body. Might he be like an Ethiopian marathon runner?) Returning to his hotel in the evening, he was not prepared for the attractive young lady in the lobby who asked if he would like her company overnight. Was this what Bob had predicted? Well, this was a time to say no, though he always found Ethiopian women beautiful.

On the first Sunday after his arrival he went to church. It was a cheerful, noisy and packed congregation of many denominations, from arm-waving Evangelists to sober Presbyterians. Sam Ross, himself a Baptist, accompanied by his wife, sidled up to welcome him to their community, having noticed him as a new face in their church. Forthwith Sam invited Gerald to stay with him and his family in their house in the leafier suburbs of Addis Ababa rather than in a hotel. Sam was a Scot who, with his family, had lived on the African continent most of his adult life. He was a hard-working gynaecologist and obstetrician in the main hospital in Addis Ababa. His wife, Morag, was a social worker. They regarded it as their mission in life to use their skills in what are now commonly referred to as developing countries. They also had a cheerful family of two young girls and a boy, all at school.

Sam picked up Gerald the following Friday on his way home from the hospital. On the way they stopped outside a shop to buy some bread and rolls. Sam asked Gerald to come into the shop with him to be given a surprise. The 'usual' order was for one hundred rolls and three loaves of bread.

As intended, Gerald was amazed and puzzled, but Sam would not explain. The following morning, a long queue of beggars and destitute people had formed outside their house. It turned out that this took place every Saturday morning. Each person got a roll, a cup of tea and a piece of fruit.

Sam also told Gerald that whenever foreign dignitaries came on official visits to Addis Ababa, the many beggars on the city's streets were rounded up by the police and taken by truck a hundred miles or so outside of Addis Ababa and left to walk back. That would give enough time for the dignitaries to go away with the impression of a vibrant city free of poverty.

Ethiopia already had several agricultural experimental stations. The principal one at Holetta in the highlands, not far from the capital, was given over to both crop production and animal husbandry. Other stations were in different parts of this diverse country – from cool highlands to tropical areas, and poor mountain range land to rich cropping areas. Facilities for breeding trials with cattle were, however, very limited and sheep improvement schemes were only an idea that had been mooted.

Sheep were a favoured source of income for farmers. This derived mostly from the meat of slaughtered, older animals but also from hides and skin. Hides and skin from both sheep and cattle were a useful export for the leather industries but they were often of inferior quality or damaged by fly strike.

The older staff in charge at the Holetta experimental station were courteous, if reserved, and somewhat suspicious of this relatively young outsider who they thought, correctly, knew nothing about their needs and conditions. But they did not know Gerald, or that his persistent yet polite questioning allowed him to learn fast. The younger staff, several of whom had had periods of study abroad, were enthusiastic and supportive from the start. This included Dagnatchew Yirgou who, though still relatively young, was a dynamic research leader at the Institute. He spoke excellent English and guided Gerald in many ways.

This is not the place to go into the detail of the problems facing the development of dairy production or the recommendations that followed. But as Gerald wrote in his report, the basic question to be answered was 'which type of cattle would be the most appropriate for a developing dairy industry, having regard to the conditions of the country'. With different climate zones, varying topography and differing amounts of available grazing and fodder, the chances were that 'one size fits all' would not be an option.

Ethiopia was home to several indigenous breeds of cattle, including zebu and sanga. Crossbreeding them to exotic breeds was an obvious first step in place of the failed importation of purebred Friesians. The choice of exotics had to be limited to representative types from among the large array of possible breeds. The Friesian, soon to become the dominant dairy breed in most developed, temperate countries, was an obvious choice. Because meat was an important by-product from dairying, the Simmental was chosen as a dual-purpose type – but really for a 'triple purpose'. Oxen of the Simmental crossbreds might prove to be better for ploughing and pulling carts than some of the other crosses. This was one of the several 'other considerations' Gerald was confronted with. A final consideration related to the economics of producing milk. Large animals need more food than smaller ones, so milk might be produced more efficiently from smaller animals, even if the total volume of milk produced was a bit less. For that reason the Jersey was the third breed chosen – it was also said, albeit not proven, that it was better for adaptation to warmer climates.

A few weeks into the project the outlines of the breeding schemes had become clear to Gerald and a meeting was called for him to present his ideas to both the research staff, who had already been involved, and to the administrators and their advisors from the ministry. It would all be new to them. The men from the ministry hoped for quick results for a minimum outlay, as they immediately made clear. The anticipated financial support from the United Nations would be a help, but they would still have to meet the costs of the facilities.

Gerald knew he might have a mountain to climb to get the breeding trials adopted. He started by reminding them of the obvious – that their beautiful country was very varied in topography, climate and richness of soil. He hoped they would agree that what may be the best animals in one area were not necessarily the best everywhere. Therefore, he proposed trialling the same breeds at several different locations. Not unexpectedly, for Gerald had heard this already, the ministry man said that each of their experimental stations liked to do their own thing. This was a problem Gerald had to face down, but first he had to make matters worse by insisting that trials at the four different stations in different parts of Ethiopia would have to be co-ordinated from Holetta. After a period of heated discussion among the representatives of the other stations they agreed that they would cooperate, provided they liked the proposed breeding plan. Gerald then set the scene for a large, long-term experiment involving several breeds. He stressed the need for large numbers of each breed, pointing out that some earlier crossbreeding trials had not been well designed. One, he reminded them, had been conducted with only ten local cows and a single Friesian bull. One bull, he told them, was not sufficiently representative of its breed. In fact, he insisted, no experiment was needed to predict that crossbred offspring from Ethiopian zebu cows and a bull of a high-yielding dairy breed would yield more milk than the pure zebu. Gerald was in the process of wearing down his audience to accept that large numbers of different native breeds, and semen from large numbers of bulls of exotic breeds, were needed to provide meaningful guidance for their national breeding policy. The final hurdle to cross was that a new manufacturing plant would possibly be needed to make liquid nitrogen for storing the semen. Some of their staff already had the skills for artificial insemination.

And so the discussion went on, reaching a final agreement that Gerald's full report would set out the details and, provided FAO also agreed, they would do their best to carry out the scheme.

For Gerald there was a mixture of wonder, excitement and fun to be had from this new experience. In the highlands of Ethiopia it was just about possible for him imagine himself back in Scotland, with the wide vistas of rolling hills and the occasional peak rising above the rest. Of course he had to replace, in his mind's eye, eucalyptus trees with birch, and acacia shrubs with gorse, as well as forget that he was several thousand feet higher up. But he could feel at home.

But it was not so in the tropical east of the country. Harar is an ancient walled city not far from Alemaya University of Agriculture, where discussions about the breeding plans were to be held. Being inside the old part of the town was like being back in the seventeenth century. In the markets it was difficult to decide what *not* to buy among the profusion of handicrafts, woven cloth or basketry.

On the second evening of their stay in the area, Dagnatchew, Gerald's constant companion, suggested a trip to the local refuse dump. This was a strange suggestion, which he would not explain. Their driver duly packed four of them into a Toyota station wagon and they set off to the outskirts of the town. A crowd of tourists was already there to watch the awesome and slightly frightening spectacle of large spotted hyenas, perhaps twenty of them, rummaging among the refuse, their large eyes reflecting the headlights from parked cars. The hyenas were attracted to the spot by men who fed scraps of meat to them and who had turned this into a nightly tourist attraction. Understandably, they hoped for a 'donation' for their troubles.

Another trip, westward this time, was again so different from the highlands. Gerald and his companions travelled to the lush, warm and fertile countryside where crops were grown for both human consumption and animal feed. This would provide one of the environments in which to compare the cattle breeds for the intended experiments, and, later, to introduce trials for sheep.

Five scientists – four Ethiopians and Gerald – had booked into a hotel made up of chalets in this semi-tropical part of the country. It was beautiful, surrounded by eucalyptus trees, acacia bushes, and grazing land. Just behind the reception building of the hotel, a large field of maize swayed gently in the breeze.

It was decided that they would go to the local town for a meal. Waiting for his companions to get ready, Gerald sat on a bench watching the sun disappear over the horizon. A loudspeaker played out rhythmic, melodious African music while cicadas chirped ceaselessly in the background. As he sat in a delighted stupor, a bedecked local woman sidled by seductively. She was very alluring and was giving the come-on in a quite subtle way. But Gerald was hesitant as his friends were about to show up. He felt guilt even from his thoughts. When he returned from town that night he half hoped to find her, but of course she was gone.

The project was accepted by the Ethiopians and enthusiastically received by the FAO staff responsible for funding. Mason complimented Gerald, saying that his report and the detailed breeding plans were the best he had seen in his time at FAO. Hugh Donald's prediction was right that Gerald's skill at dotting the i's and crossing the t's would impress.

Chapter 15

Melbourne Conference and a Return to Ethiopia

Before returning to Ethiopia to find out what progress had been made in the cattle experiment, Gerald had the good fortune to be given the time to attend a World Conference on Animal Production in Melbourne, Australia, in May 1973. He travelled with a party from the British Society of Animal Production, most of whom he knew well due to his long association with the society as editor, and at various times as a member of its council and of its publications committee, including time as chairman of that committee. The society had arranged for a week's tour of New Zealand before the Congress.

It was Gerald's first visit to that country. He was enchanted by its natural beauty and by what seemed a more relaxed lifestyle. Furthermore, animal production was an important part of the economy and therefore held in high regard. He wished it were so in the UK, where animal production was of interest only in the rural community whilst most of the population cared little.

The New Zealand tour crammed a lot into one week. The group visited agricultural research stations and universities, many of whose staff had spent time at ABRO and the Institute of Animal Genetics or studied in Edinburgh. It was a bit like visiting friends. Tourist highlights, such as the thermal region of Rotorua and the nearby trout sanctuary at Rainbow Springs, were not neglected.

An overnight stay at the Hermitage in the Mount Cook National Park and sight of the majestic mountains was something Gerald never forgot. But curiously, what he remembers with the greatest affection from the trip was the wonderful dessert – the pavlova prepared by women of rural institutes who entertained them on visits to farms. The women held competitions for the best pavlova, but to Gerald they all tasted delicious, and it has remained his favourite dessert to this day.

The conference in Melbourne was intense, with a multitude of science talks during morning and afternoon sessions and with workshops for special discussions. Social functions during that week were restricted to just the evening welcoming party, which gave an opportunity to renew acquaintances or friendships with delegates from other countries, and then the obligatory conference dinner towards the end of the week. The science and indeed his own talk to the conference have long faded from Gerald's memory, but what he recalls vividly is a lapse in his behaviour. Following the conference dinner he had gone to a near all-night party. The next day he fell fast asleep the moment he sat down in the conference hall, snoozing throughout the entire morning session. He feels ashamed of that to this day, though no one confronted him about his misdemeanour. Perhaps there had been others in the same boat.

Gerald returned to Ethiopia in 1975 to review the cattle- and sheep-breeding experiments that had been set in motion – or not.

The atmosphere in the country had changed. The Emperor had been deposed and a military, communist-style junta had taken over the government. Foreigners were no longer very welcome. Sam's family had left for South Africa, but Sam himself felt it his calling to stay on for the sake of his female patients. He had given up his house; no more rolls and tea for beggars, no more bible study groups of young people. Sam had moved to a room in the living quarters of the hospital. Gerald once again stayed in the Hotel Ethiopia. Most of the girls plying their trade had been evicted from the streets around the hotel area, and the small neighbouring hotel where the girls used to take their clients had been closed. All to the good, no doubt, but the relaxed and cheery atmosphere of earlier times seemed to have gone.

The research programme with cattle had started, but encountered some problems mainly through underfunding of facilities and insufficient technical staff. The liquid nitrogen plant, essential for storing semen, had also broken down. Most of the effort now had to go into suggesting, in detail, the alternatives for overcoming the problems. The sheep work had not even been started.

Travel by foreigners was restricted by tighter controls from the new government. Most of the 'review' that Gerald had been charged with took place, therefore, in and around Addis Ababa and at the nearby experimental station of Holetta. There were no visits to the enchanting lands of the east and west.

During the second week of his stay, Gerald became unwell with a chest infection. Sam, still his friend and 'guardian angel', diagnosed pleurisy and insisted that Gerald move from the hotel to the leper colony, supervised by his doctor brother, on the outskirts of the city. There Gerald would be cared for and allowed to recuperate. Gerald never met or saw any lepers as he was housed in a separate chalet.

His recovery was hastened by the loving care he received, which was just as well as he was due to take the chair at a pan-African conference on animal production to be held in the imposing Africa Hall, a prestigious building and United Nations conference venue. He never quite found out why he was chosen for this daunting task, but assumed that, at that time, he was the most senior FAO consultant in the area – and the only one concerned with animal production. Talks on animal production projects and schemes were given by senior scientists and administrators from several of the

African countries to an audience of several hundred specialists and administrators. Question time followed the formal presentations, with the speakers assembled alongside the chairman on the platform. Suddenly the studied atmosphere of the conference was broken. Gerald recalled what happened exactly.

A voice from the hall shouted out, 'Mr Chairman, I wish to draw attention to and protest about the very unequal distribution of aid by the people in authority in Rome . . .'

A murmur went through the hall as a string of accusations was called out. The speaker was, it seemed, well known to his audience. What was Gerald, as chairman, to do? He was a foreigner among many of the African elite in the field of animal production. But this was an intrusion to the flow and purpose of the conference. Gerald hoped he could stop it. He rose to his feet. He told the speaker that he regarded that as a political question that had no place at a scientific meeting.

After a momentary hesitation the speaker resumed his seat, to Gerald's relief. At the end of the proceedings Gerald had hoped to escape back to his hotel but was waylaid by his 'political' friend. He claimed that he was prevented from making an important point about aid, but accepted the chairman's right to stop him. He added that he had no hard feelings.

Before Gerald left on his way back home via Rome, two of his Ethiopian friends asked him to come with them to visit Dimaketch, a pretty girl he had met as the housekeeper and friend of a Scot from Edinburgh, a long-term, resident consultant for poultry production in Ethiopia and neighbouring countries. He also had left the country with many other foreigners following the change of government. Dimaketch, however, now had a baby. As was not uncommon, the father of the child, a friend of the Scot, had a wife back home, but fortunately he had accepted financial responsibility for his child and its mother. Dimaketch hardly remembered Gerald, who had been only a sporadic visitor to their house, but now she revelled in her newfound status. It was only a room and small kitchen, but it was hers. She languished in a large bed nursing her baby and surrounded by large pillows as her visitors entered. Now she could show off. She had a servant woman, much older than herself, crouched on the doorstep. 'Light me a cigarette,' Dimaketch told the old woman. Duly lit for her, she smoked ostentatiously in her bed. It was another introduction to a different culture.

As a postscript, it is worth noting that the cattle project did proceed almost as planned. It was greatly assisted by the appointment, by FAO, of a brilliant young Egyptian animal geneticist, Salah Galal, as a consultant – ostensibly to get the recommended but neglected sheep trials under way. However, he also rescued the project from potential disaster following the breakdown of the liquid nitrogen plant. He arranged somehow for sufficient bulls of each breed to be found to fill the gap. He also brought his statistical and computer expertise, honed at Ames, Iowa, to bear on the analysis of the data from the cattle project. Thus the project started to provide material for many scientific publications by young Ethiopian scientists for many years to come.

Dagnatchew stayed in touch with Gerald long after the end of his visits to Ethiopia. He assured Gerald of the influence the cattle experiments had had on the breeding policies pursued in his country. In particular, he wrote that the introduction of the Jersey breed had had a marked effect on crossbreeding programmes with local breeds. Because of its relatively small size its crosses had more modest feed requirements than the crosses with the other two exotic breeds. As Gerald was to discover in later years, such a positive impact from one of his missions was rare indeed.

Gerald and his family were later to relive some of the beauty and splendour of Ethiopia, including parts of the country he had been prevented from visiting by the travel restrictions on foreigners. The

BBC made an excellent documentary series about the country – Gerald and Sheila rented their first colour television set to watch it and never went back to using their small black-and-white set.

CHAPTER 16

A Homecoming

Those people who knew Gerald well, especially his colleagues, were impressed by his ability to immerse himself in a project. Along with his quick mind, his chief quality was that of persistence. He worried at a topic, overcoming each new problem as it arose. For this reason, he was well suited to his chosen career. The close study of animals, and the devising of relevant experiments, thrilled him, nourishing his almost boyish intellectual curiosity.

He has often said that a career in science can be a rewarding and highly privileged life. Most researchers have the satisfaction of pursuing ideas that really interest them. Few other professions, apart perhaps from artists and writers, are given that freedom. In addition, for many there is the pleasure of recognition, and of attending national and international conferences where they can meet their peers. For Gerald there were many such occasions. These were meetings concerned with animal production and breeding and of course those related to trace element metabolism.

It was at the time of a trace element conference in 1981 in Perth, Western Australia, that Gerald became reacquainted with the Carter family. They were descendants of people from Ireland who had been transported to Australia in the nineteenth century. Gerald's own farming experience had been with what one might call the average British mixed farm – a few hundred acres, some cattle, sheep and pigs. Even the largest of ABRO's hill farms did not prepare him for the 200,000 acres and more than 20,000 sheep that made up the sheep station owned by Ron and Val Carter in Western Australia. He had first met these Australians at the animal production conference in Melbourne some years earlier and now they had invited him to spend a week with them in their home.

Ron Carter and his wife, Val, had won the right to farm 200,000 acres of virgin land. That had meant firstly vying with others to prove to the government that they could make the best use of a particular parcel of land. There was no house and no water. They had to find a place to dig a well and they had to build their own house from scratch. It is there they brought up a fine family. Their sheep were of a high-class strain of fine-wool Merino and Ron had become a well-known breeder. What tales they had to tell! For Gerald it was an insight into a different kind of life from his and he learned a lot. One evening, when with Ron and Val in a neighbouring town, Ron asked Gerald to drive their four-by-four vehicle back to their property, as they wanted to stay longer to talk with friends. Not wanting to admit that he had never driven such a massive vehicle before, and especially not at night, Gerald replied that it would be 'no problem'.

He arrived haltingly at the long farm driveway, and had to brake abruptly as, shocked and scared, he was confronted by a very large kangaroo, caught in the headlights of the car. 'Go away!' he whispered as he viewed the seven-foot animal. And mercifully, it hopped away out of sight.

After his adventures down under, Gerald arrived back in Edinburgh with a mind full of reminiscences of all the people he had met. He alighted from his taxi at Dalhousie Terrace, tired but happy. The conference had enjoyed his paper, which provided yet more new evidence on genetic involvement in the absorption by sheep of copper from their diet. His friends told him that he would be famous – but he took that as a tease.

The driver of his taxi carried his luggage to the door of his terrace house. He pushed open the door to see Cindy, his mongrel collie, sitting on the stairs that faced the heavy, broad front door. As he later told some of his friends, it was the dog that alerted him to the seriousness of the situation. The black and white of her shaggy coat looked flat, her head was low and she was quite crestfallen and unhappy looking. Her tail stayed down, and she slouched down the stairs to greet him.

Gerald patted and made a fuss of her, and he followed when she turned and made her way into the kitchen. On the table was an envelope, white on the stark wooden table. Cindy lay down as he slowly opened it. He read his wife's small handwriting.

> Now that the children are away, there is no point in my staying here any longer. I have taken most of my things. I may have to come back if I have forgotten anything. Then I will return my keys to you. Sheila.

Dazed, Gerald slumped down on a chair, his coat still on, his face ashen. He looked around the kitchen, trying to take in the situation. Cindy lay down at his feet.

For several minutes he remained immobile. Cindy must have seen the suitcases, he thought, and worked out that, like the two teenagers who lived here before, Sheila was moving out. Maybe she had said goodbye to the dog. More than she said to me, he thought.

Some of Gerald's close friends knew of the many turbulent times in the marriage. It was known by a few that it had been the children that, in recent years, had kept the two of them together.

Both children came home, and it turned out that Sheila had let them know about her decision to split from their father. She said that she had long been unhappy, and that she would be better to make a new home for herself in Edinburgh. This had taken place while he had been in Australia. On the day of his homecoming, they knew their father was in for a shock, and had arranged to be there to console him. There they stood, fresh and healthy, maybe upset, maybe perturbed at this happening. Andrew had come from his student flat in Edinburgh, where he was studying medicine at the university, and Ruth had travelled up from Doncaster Art College.

As Gerald told his friend Alan Dickinson, the children were a great comfort to him at this emotional time. Andrew said that he would be home after his exams, and that things would settle down. They both said that they would be home regularly and that things would be all right. The two of them had brought food to make a meal together that evening, and they sat down and talked out the situation. Soon, dumbfounded, he watched them rise and prepare to go. With heartfelt goodbyes, they left for their various affairs and activities. They were tactful, but he could see that they were relieved to escape. In their blue jeans and scarves they closed the large front door, and were off down the street. He remembers watching them from the bay window as they nodded and talked together. They had said they had busy evenings ahead.

And that night when he went to bed, Cindy slunk into the bedroom and lay down silently by his bedside. She had never done this before.

The next day, he had to appear back at his research institute at King's Buildings to give a report to the staff about his trip to the conference in Perth, Western Australia. They listened attentively and the director of the institute congratulated him on his first-class work. They all welcomed him back, but Gerald cut the meeting short, drove to the field laboratory in Roslin where he also had an office, reached his own room and shut the door. He could tell them of the split with his wife later.

He knew that several of his colleagues were already divorced or separated from their wives. It's this job, he thought. It's too hard for relationships. Hours and hours of experimentation, of reading, of studying, of writing scientific papers. It was a matter of complete absorption in your subject. Often home life had to take a back seat. And the scientific papers that had to be produced, the trips abroad, the conferences – they were all exciting and absorbing, but not for the wife and family.

But he knew that this was only half the story. He hadn't got on with his wife for years. There was a clash in their personalities, and in their careers. She was a very clever and intelligent woman, unhappy at her situation, and perhaps somewhat jealous of her successful husband, and then of course there were her depressions.

'Alone!' he whispered to himself, cloistered in his little office.

Chapter 17

Edinburgh Years and a Fateful Meeting

Some time after Sheila left, Gerald and she were divorced. In the years following the break-up, Gerald became even more absorbed in his work. When not concerned with his experiments, writing scientific papers, lecturing or attending lectures and going to conferences, he found great solace in attending the Congregational church at the corner of Bruntsfield Place and Chamberlain Road in Edinburgh. As time wore on, the congregations of his church and of the neighbouring Church of Scotland, at the so-called 'Holy Corner' in Edinburgh, merged into one church. Gerald, along with the other members of the council of the Congregational church, were ordained church elders of the Church of Scotland for the newly named 'Morningside United Church' – wits always suggested that 'Morningside United' sounded more like a football club.

In the run-up to the merger Gerald became a leading figure in plans and discussions on how to turn a large Victorian church building, the former Morningside Church of Scotland, into a community centre when no longer used for worship. A steering committee was set up with members from the four denominations then with churches at Holy Corner (viz. Congregational, Church of Scotland, Episcopal Church of Scotland and Baptist). By popular choice, he was voted in as chairman of the steering committee for the project when, after only a few months, its first chairman, Jeff Maxwell, left Edinburgh to take up his appointment as director of the Macaulay Institute in Aberdeen. After the Holy Corner Church Centre was founded, Gerald was made chairman of its governing council and remained as such for six more years.

At home in Morningside, life was quiet and a little lonely. Gerald's son, Andrew, and daughter, Ruth, then in Aberdeen, were both busy with their studies, and his only companion was the dog. They became ever more attached to each other. For a few months, Andrew returned home from his student flat to keep his father company. But the lifestyle of the young man did not fit in easily with his father's attempt at being cook and housekeeper. Though the two had great affection for each other, it seemed best on all sides that Andrew should return to stay with his mates, while Gerald stayed at home, with only occasional visits from friends to break the loneliness.

Sometimes Gerald would take a walk in the beautiful Edinburgh Botanic Gardens. He would reflect on the week past and the coming weekend visit from his mother. These visits were half welcomed, half dreaded, as Luise would admonish him for his untidiness and for not eating enough. She had retired now and had bought herself a flat just a short bus ride from her son's house in Dalhousie Terrace. She tried always to see something of him on Saturdays. Luise still felt she could interfere and advise Gerald on how to run his life. She had even obtained a key to his house by a ruse. Meeting her granddaughter, Ruth, in town one day she had claimed to have lost her keys to Gerald's house and asked to borrow Ruth's – upon which she promptly had them copied. Secretly, mother and son cared deeply for each other, but often they were found arguing heatedly on the phone. It was a kind of continental formula that they had developed.

It was after one such walk in the Botanics that Andrew phoned and Gerald told him of his meeting with a pleasant lady named Margaret. It was a lovely, sunny day, and the woman and her daughter were seated in the park café discussing where they might find a place locally for an evening meal. Gerald had sat down at a neighbouring table. He leaned over and told them about the Chinese restaurant he sometimes went to, Loon Fung, on the other side of the park.

In the end they walked there together and Gerald went inside with them. He was greeted by the proprietor with a great show of friendship, and was persuaded to join his two new friends for dinner. They had exchanged phone numbers when they left, suggesting that they might do this again.

They did meet again, and he took her to an orchestral concert in the Usher Hall. But after that the budding affair languished. Margaret seemed to have met someone else. Months passed. Once, when Gerald was passing through Biggar, he stopped outside her little cottage. But he heard a male voice over the garden wall, and he hadn't the nerve to go inside. So he drove off home. He wrote a letter to her, friendly and conversational, telling her about his divorce, and about his two student children. She wrote back saying she was now a grandmother. Shonagh, a little girl born to Laura, Margaret's older daughter, was now months old. Also her younger daughter, Maggie, had been chosen to be Cinderella in the village pantomime, and they had acquired a little dog, a Springer Spaniel.

Then came a strange development. Gerald was invited to Greece by his colleague, Ian Mason, and Ian's wife, Elizabeth, who had a little fisherman's cottage on the island of Aegina. On arrival, Gerald relaxed in the sunshine of the beautiful place, the house being just yards from the sea. An unsettling part of the holiday with these somewhat eccentric friends was that they sunbathed on the roof of their little home in the nude, and expected him to do the same. Then a female friend of theirs from the UK arrived – a similar arty, actressy type as Elizabeth. It transpired that he and this other guest, a lady called Joy, were expected to share a room for sleeping – one of only two bedrooms in the cottage. They did so with great embarrassment though the beds were far apart. When

Gerald asked Ian why they could not have split up and had the two males in one room and Elizabeth and Joy in the other instead, he retorted that the arrangement was Elizabeth's wish.

However, there was an unexpected consequence to this holiday. Joy had once lived in Biggar and had been principal of the Biggar Theatre Workshop for some time. She knew Margaret and her daughter, Maggie, well. Margaret and Joy had both been widowed, which gave the two a kind of bond. When Joy got back home she took her photograph album round to Margaret's cottage to show her the little house she had also bought in Aegina, and to tell her that Gerald Wiener had asked after her health and more.

At Christmas, Margaret phoned. She had indeed being going out with someone else, but this had fallen through. She had to steel herself to phone him after so many months. She explained that she had heard from Joy, and all about his holiday in Aegina. Then she dropped a bombshell. She asked him what he was doing for Christmas, and if he would like to have dinner with them in Biggar. Gerald was startled, and delighted to accept.

And that was the start of a friendship, and some fun, eating out, visiting other friends, and going out to films and the occasional theatre. One night, Gerald threw a dinner party at his house in Dalhousie Terrace and invited the Masons and Joy Graham-Marr. Margaret was helping with the cooking – Gerald was making coq au vin. The three guests arrived at the door of Gerald's house. A loud, confident voice was heard. It was Elizabeth Mason exclaiming, 'Here I am, encased in mink from head to toe. We've just been to the Kelvingrove Museum in Glasgow – Joy, Ian and me. It was a stupendous exhibition. The Post-Impressionists! Just wonderful!' This was typical of Elizabeth, and a bit overwhelming for Margaret, who did not know Elizabeth's ways. However, the party was a success.

Gerald was keen to introduce Margaret to his friends at the Holy Corner churches. He wanted to show her the work being done, and the discussions going on about the vexed subject of fundraising. The plan of the steering group, of which Gerald was chairman, was to convert the enormous, empty Victorian church building at one corner of the crossroads on Morningside Road into a centre to be used by all groups of the community – mothers and toddlers, disabled, discussion groups, students from nearby Napier College (now University) and so on. They also wanted to start a café, open to all residents and to passers-by. It had been estimated they would need to raise a million pounds if they were to follow the ambitious plans drawn up by the architect who had won the architectural competition mounted for this purpose.

One Sunday, Margaret was persuaded to attend church with Gerald. Stuart Miller, the minister, mixed with the congregation over coffee and Gerald introduced Margaret to him. He was a reserved, grave, intellectual man. People found it hard to talk to him, but Margaret and he got on quite well in conversation. He mentioned the million pounds that were needed and said that they did not know how to raise that kind of money.

Margaret threw in the suggestion that they should try to identify someone famous who had been connected to one of the four churches in the past. They had been in existence for a long time so she felt there must be a few such people, maybe in other parts of the world. Perhaps a rich American, for example.

Stuart said that the only one he knew of was Eric Liddell, the famous runner. He said that he and his family had been members of the Congregational church at Holy Corner and in fact, he had taught Sunday school there.

When Margaret told him, Gerald jumped at the idea. After visiting Eric Liddell's sister, Jenny, in Edinburgh, accompanied by Noel Littlefair, one of his colleagues from the council with a special responsibility for the building, and writing to Eric's daughters in Canada, he received their enthusiastic support. Gerald then proposed the new name to the centre's council. It was welcomed and adopted and that is what it is today, the Eric Liddell Centre. What is more, it helped Gerald's successors to raise the large sums required for the major conversion of the building. It was no doubt helped by the wide popularity of the film *Chariots of Fire*, based on the life of Eric Liddell. It had made Eric's name and Olympic success so much more widely known. A happy set of coincidences all round.

Chapter 18

Wedding Bells

Margaret owned a picturesque little cottage, nestled behind the main street of the little market town of Biggar, just over half an hour's drive from Edinburgh. The rooms were small, the place was easily furnished and heated, and Margaret lived there with her unmarried daughter, Maggie. There was a lovely little garden planted by the previous owner with only blue flowering plants. Margaret and Maggie would sit there after their evening meal on summer nights and look out on the gentle hills of the Southern Uplands.

The friendship with Gerald Wiener, which had become a permanent part of both their lives, proceeded, and on this evening Margaret had returned home from Gerald's house after ten in the evening. The telephone rang. It was Gerald phoning to say that she had left her jacket in his house. It was a fashionable mink jacket that she had bought for her elder daughter's wedding, but she told him to hang it up and she would collect it in a few days. As she was putting out the lights, the bold Gerald arrived at the door. It was after midnight, but Gerald looked delighted with himself and his daring in driving so fast from Edinburgh. 'Just because the lady loves her mink!' he said as he handed it over. She was to see a new side to the serious-minded scientist.

It was clear that they were becoming used to each other. Both of them had been through the mill of life. Gerald had been through the trauma of divorce, and Margaret had been widowed after the sadness of her first husband's many years of illness and death. They had developed their own survival skills, but it seemed that they were slowly falling for each other. As they say, a small triumph of hope over experience.

In June that year, Margaret had to attend a weekend workshop course for head teachers of the area. The event had been arranged at a big country house in Ayrshire used by the county council and also at times for educational purposes. While there, Margaret was called to the phone. Several other head teachers, her friends, were with her and wanted to find out what the call could be about. When she put down the phone, one of them asked why her face was so flushed. She explained that her male friend had phoned to say that he had bought a seven-roomed house in Biggar, and that he said she would now have to marry him.

In October 1985, Gerald and Margaret married. The wedding itself was held in Biggar Kirk among family and friends, and officiated over by Rev. MacNeave, Gerald's friend of many years from the former Congregational church in Morningside and one-time missionary in China. How appropriate that the best man was Jimmy Harris, manager of Blythbank Farm. It was he who had overseen Gerald's long-term sheep experiment there from the start and had over the years also become a good personal friend. The wedding reception was small by the standards of the young nowadays but it included all the family, from the youngest baby grandchild to the oldest, aged two, and also both the bride's and groom's mothers and long-time friend Mrs More. The honeymoon, in Paris, was quite short as it was restricted to Margaret's school's half-term holiday. The couple arrived at a highly recommended hotel near the Paris opera house and not far from many of the other famous Paris landmarks. Their room was not what was expected. It was small and dingy; the window looked out on the wall of a block of flats not more than two yards distant and it stank of cigarette smoke.

Gerald complained about the poor quality of the room and the smell of tobacco. He was angry and told the hotelier that he had specifically asked for a superior room as it was for his honeymoon. They apologised, saying that they had no note of such a message, but next day they were given a large, sunny room facing the front of the hotel with period furniture. This was very pleasant, and the couple were delighted.

It was at this time that Margaret discovered Gerald was an inveterate explorer of cities and they set out on the first of several long treks across Paris, with stops at galleries (the Louvre, of course) churches and, to please Margaret, department stores. On their way through Montmartre Margaret spotted the beckoning sign of the Moulin Rouge. She pleaded with Gerald to take her there, and throwing all caution to the winds, Gerald splurged out on the expensive tickets for the show.

And so they enjoyed that lively evening show preceded by a dinner, with a bottle of champagne per couple. They shared a table with a Russian couple who, in halting English and with much laughter, confided that they held four jobs between them and had made lots of money.

The end of the holiday came all too soon. The little cottage in Biggar and the terraced house in Dalhousie Terrace in Edinburgh were sold and the couple set up house together in the Old Schoolhouse, Biggar. During the week, Margaret travelled to Lanark to her school, and Gerald travelled to Roslin or Edinburgh. The following year, Gerald, being now sixty years old, had to follow the civil service rules and retire from his position in ABRO. This did not please him at all, and for years he travelled back and forwards to the agricultural complex of offices and laboratories at Roslin where he continued to do work that interested him.

Gerald's retirement dinner was a friendly and happy affair, although Margaret did not understand many of the in-jokes of the sometimes rowdy company. At one point, one of the party did an imitation of Gerald arriving with his coffee to sit down, his teaspoon still in the cup. He held on to his own spoon because there was a dearth of spoons in the staff room, and he wasn't losing his one. At one point the company asked Margaret to say a few words, but she, being a bit overwhelmed by the roomful of colleagues and friends, who all had so much in common, declined to make a speech.

It was clear that his colleagues thought highly of him. Gerald had been, after all, many years on the research scene in Edinburgh, and was the longest-serving member of ABRO. He had helped in the appointment of many of the younger staff and took special pleasure in the fact that some had been members of his department of physiological genetics. This included Roger Land, who became director in Gerald's final years before retirement, and also Ian Wilmut, who was to achieve fame, and a knighthood, as the leader of the team that developed the first mammal in the world cloned from an adult cell – Dolly, the sheep.

Chapter 19

Return to the USA

Rudi made crêpes Suzette that first morning they spent in the San Francisco Bay Area in 1992. The partner of Gerald's aunt Thea, he was a short, broad-shouldered, good-looking, non-Jewish German in his sixties. He told Gerald and Margaret the story of how he had jumped ship from his German merchant vessel in the dying days of the war. He had found work easily, and after he married an American woman the authorities had eventually stopped bothering him.

In later life he divorced, and met Thea when she came to work at his delicatessen. They were very happy together and had recently retired. Their apartment in San Mateo was very comfortable. In the living room there were two large television sets, side by side. One carried programmes in German, it seemed mostly with 'oompah' bands, the musicians all dressed in traditional costumes; the other was for English-language programmes, and seemed to show nothing but quiz shows. These the old couple watched avidly.

The building had an underground communal garage, a swimming pool for the use of the residents, and two caretakers who looked after the hallways and lobbies. Gerald and Margaret had Rudi's room for a few days before setting out on their journey across America. Vera, Thea's daughter, joined them for lunch – chicken barbecued by Rudi on the balcony of the flat, which overlooked San Mateo.

Thea was quite emotional at meeting her late brother's lost child again. She thought he was even more like Paul, his father, this time. It seemed to her that Paul had almost come alive again.

She brought out an old photograph album. According to Thea, Paul had had many girlfriends, and at least three wives. The first was Luise, Gerald's mother, followed by a fiery Hungarian woman, who was part of the reason for Gerald's parents' divorce. The Hungarian left him when he was taken to the concentration camp just before the war. The album was beautifully decorated with coloured frames to the old photographs. Paul seemed debonair, sometimes pictured playing the accordion with his four-piece band, other times with girlfriends in the great cities of pre-war Europe. The album was presented to Gerald to keep. He now had two albums of his father's: the one with photographs of him as a baby and the later one of his father's own life.

The next day, they hired a car and drove out to lunch at Half Moon Bay with Thea and Rudi. Though a bit nervous of driving in San Francisco, Gerald valiantly drove over the bridge down to San Ramone to visit Judy and Tom, Vera's daughter and son-in-law, and Lorien, their talented teenage daughter. After a great meal of lasagne and salad, they drove up the local mountain road to Mount Diablo, in the state park of that name, to see the panoramic view from the top. The evening was spent admiring Tom's collection of old comics and general bric-a-brac collected from garage sales over the years. So extensive was his collection that Judy had made him build an extension to their little detached villa to house it all.

Gerald also met his second half-brother, Pete, for the first time. Pete was married to a woman also named Judie. At that time they had two teenage 'all-American' boys. In their lovely house they had a rumpus room where there was little furniture, just a carpeted floor with floor cushions and an enormous television set. There the three males of the family watched American football. The family went in for big equipment including a cavernous refrigerator, larger than any Gerald had seen before. They also had an enormous fifties-style jukebox in a room where otherwise there was little furniture. On display as well was the accordion that had been their father's prized possession, Pete said.

The visitors from UK were made very welcome, and the food was delicious. They sat down at a table where there were cold meats and salads of every description – pastrami, salami and turkey were laid out, rolls and seed bread, and endless glasses of Californian wine. The road back to Judy and Tom Reid's house in San Ramone was discussed and it was decided that they should leave before seven o'clock, so as not to make it difficult to find the exits and interchanges of the highway in the dark. In the garage as they left, Pete proudly showed them his 1974 sports car, number plate PWS TOY, meaning Pete Wiener's Toy. A very American couple, they stood in front of their house, Judie with her inch-long, pink-painted fingernails, her eyes bright beneath her pink eye shadow, and Pete wreathed in smiles as they waved goodbye to their guests.

The next day, back in San Francisco, Gerald and Margaret found a crowd gathered down by the harbour. Amidst the noise of the men as they cleaned up the decks of their boats, and the slap of the seawater, was a close gathering of at least fifty enormous sea lions. The British travellers were enthralled.

The couple from little old Scotland found interest and excitement enough to last for many an hour. But they had an appointment. They were meeting with Gerald's cousin Vera and her husband, Hans. Within ten minutes they were being driven up the hill in Hans's car, to the top of Lombard Street. This street is famous for its steep gradient, involving many sharp twists and turns through flowerbeds and pretty pavements to land again at the front of the great Bay Harbour. They went on across a bridge to a busy seafood restaurant where some of the Wiener family were awaiting them.

Everyone had dressed up and it was easy to see that this was to be an occasion, a special family meal. As they entered through the glass-and-silver doors of the famous Spinnaker restaurant, built partly on a promontory reaching out into the sea, their eyes were drawn to the smartly dressed waiters dashing around. The tables were covered in stiff white tablecloths and wine glasses sparkled under the lights. There were windows on all sides displaying views of the great Pacific Ocean and, in the distance, the great Golden Gate Bridge.

Chapter 20

The American Family's Story

Hans was the host at the celebration meal that marvellous day in the Spinnaker restaurant. As he greeted everyone, Vera stood by him, quiet and happy. Vera is a serene, grounded lady, and a very skilled amateur painter. It is said that when she and her family first arrived in the States, she was the one who found work, being able to speak English. Her grandparents had arrived too, and the young girl's valuable earrings were sold to keep the whole family for the first few months.

But Hans was the party animal. A tall, rangy man, he loved people, and people loved him. Gerald and Margaret were immediately drawn to him. As Jerry, Gerald's brother, said, 'Yeah, Hans is a lot of fun to be around.' He had the party seated at a great round table, menus were given out and everyone was told to order anything that they wanted. Margaret, tempted by Vera's order of Beefeater gin on the rocks, opted to have the same.

A large part of what was offered was fish or seafood. The place was crowded with people, and enticing plates of marvellous food were passing all the time. Gerald and Margaret were taken aback by this opulence. The food was superb, and when they were relaxed over coffee, Hans started the talking, as Gerald recalls.

'It is now more than forty years since we came to this wonderful city. You know, Gerald, we all love it here. There was a time when they tried to move us to Texas. But we rebelled, and absolutely refused to be moved.' Thea joined in, explaining that she told the officials that they could not go to Texas because her father could not ride a horse. That made the officials laugh, and so they got to live in San Francisco, where they wished to be.

Hans told of how he worked in a sweatshop for eleven years. Determined to get out of that job, he went to night school, and learned the dry-cleaning business, following which they managed to set-up their own dry-cleaning business. He laughed, and threw back his whisky. 'Now I have to work harder than in that sweatshop!'

They sat around talking about the days before the war. Hans had lived in Breslau. His father was born in the Ukraine, and had fought for Russia against the Germans. He was captured by the Germans and imprisoned. On his release, he decided to stay on in Germany, where he built up a tailoring business. Hans's family, like Gerald and his mother, started to realise that the political situation for Jews was becoming very bad and decided to get out of Germany.

Gradually the full stories of the Wieners, Paul and Ursel, Thea and her husband, and Vera and Hans Gelfand and his family were explained to Gerald and Margaret. It seemed that they, and many thousands of other people in peril in Germany, escaped from that country and found the only place that they could escape to was Shanghai in China. No visas for entry were required, unlike most of the rest of the world.

Arrival in Shanghai was a complete culture shock for the Jewish refugees. The disciplined behaviour of pedestrians on pavements and of vehicles on roads that they had been used to became a distant memory. This was China in 1939, and Shanghai was a chaotic place where parts of the city were being administrated by various nationalities including the British and the French. There were Russian residents there too, and eventually as the war progressed there were Japanese soldiers everywhere.

Hans called it a God-forsaken place, saying it was hot as hell in the summer, and teeming with people of all nationalities. The Jews soon had a school built, and that is where Hans and Vera met. Vera was a clever, hard-working girl, and picked up the English language very quickly. But Hans hated school and, as he said, he played around a lot. At this admission from Hans, Vera exclaimed that he hadn't changed much as he got older.

Hans protested when Vera jokingly ran him down. He turned to Gerald, saying, 'Don't listen to her, she was brought up by a dog, that is why she barks at me all the time.' Vera shook her head, explaining that the dog was the family pet they had in Germany at that time. The dog used to sleep in her bedroom when she was a baby in her cot. She smiled as she told them that the dog stayed with her until she woke up, at which point it trotted downstairs to tell her mother.

By this time Thea had finished her meal, and she joined in the chat. Her tone was more serious when she spoke, telling her nephew that he had been lucky in his escape from the horrors of the Nazis. Gerald recalls clearly how Thea leaned closer towards him to ask if he knew that his father had been taken to a concentration camp just before the start of the war. He did not know this, for his mother had kept this information from him. He remembers the feeling of sadness these words brought him at the time, and he felt a churning in his stomach at the whole idea.

As the day wore on, Vera suggested that they all went back to her house a little outside San Francisco to have some coffee and to look at photographs. Gerald was delighted at this invitation. He wanted to speak to Thea and Vera to find out more about his father.

The Gelfands' house was at the top of a hill that looked over San Francisco International Airport. In their back garden they had a small swimming pool, which glistened in the sunshine amid luxurious seating and lounge areas. As the afternoon wore on, amid the photographs and reminiscences, Gerald had a new sensation of being at home among his family. It secretly brought tears to his eyes. When it was time to leave, he asked Vera if he and Margaret could meet her and perhaps his aunt again, so that they could talk further. He explained that he wanted to write down some things, so that he would remember them when he got home. He also carried a small recording device for dictation. It was arranged to meet again two days later

They suggested a restaurant in Chinatown. The food was very authentic Chinese with sensational dim sum snacks being constantly offered to them. After a while Thea started talking only to Gerald, from where she sat beside him at the large round table. She wanted to speak to him about his father, and of how she had visited the camp where he was detained, and of how she had spoken to the superintendent at that camp, and of how she had pleaded with him. The man was a promoted soldier of the German army. Her words still stay in Gerald's mind: 'My brother has done no harm. Why have you detained him here? He is an artist, a pianist. He is an accomplished musician.' She spoke that day of her own distressed state as she tried to get her brother out of the camp.

Gerald shook his head, feeling some of her emotion. She went on to say that the captain, or whatever he was, told her that Paul Wiener had been caught crossing the border into Germany with a lot of money, and that his papers were not right. She said that she became angry at him saying this, and that she told him that of course her brother would have money because he went abroad to make music with his orchestra expressly to earn money. She explained to the superintendent that her brother was all the family had to earn money, and that the money was to keep her and his mother and father.

She then said to Gerald that she pleaded so much with the man that she wore him down until he gave in. He said that if she could come there in the next week or so, and show him a ticket for her brother to leave Germany, then he would let him go. Thea went the

Top. Degree and non-degree Agriculture students with Joe Gordon, acting director of studies, at George Square, Edinburgh, 1947 (Gerald in back row, circled).

Above. Reunion party with Prof. Peter Wilson (back row, extreme left) at King's Buildings, 1997.

Left. Dolly the sheep.

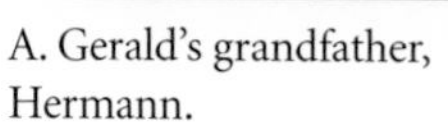

A. Gerald's grandfather, Hermann.

B. Gerald's mother, Luise, aged thirty-six.

C. Aunt Erna with Marion, aged nine.

D. Cousin Marion and Gerald, aged five.

E. Uncle Erich.

Gerald, aged three.

From the right: Thea, Paul (Gerald's father), Luise, Paul's parents and relations, 1925.

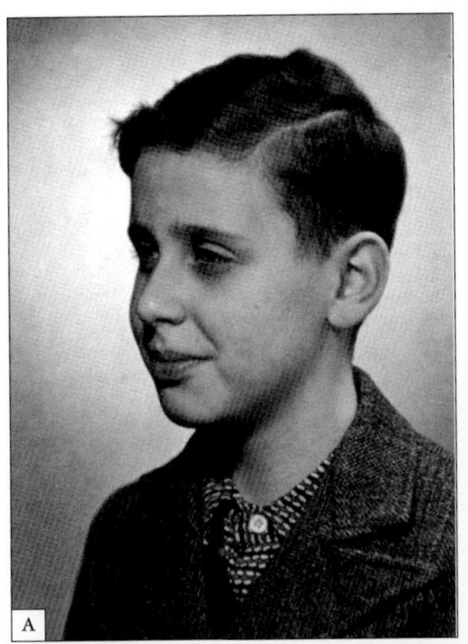

A. Gerald, aged twelve.

B. St Mary's (University) Church, Oxford. (Mieneke Andeweg-van Rijn/Alamy Stock Photo)

C. Gerald's first graduation (BSc, Edinburgh), 1947.

D. The Spooners: Rosemary (left) and Ruth (right).

E. Gerald with best friend, Hardy Seidel, 1980.

The three brothers (from left): Gerald, Pete and Jerry.

Some of the American family. Aunt Thea is seated front row, centre.

Luise, aged eighteen.

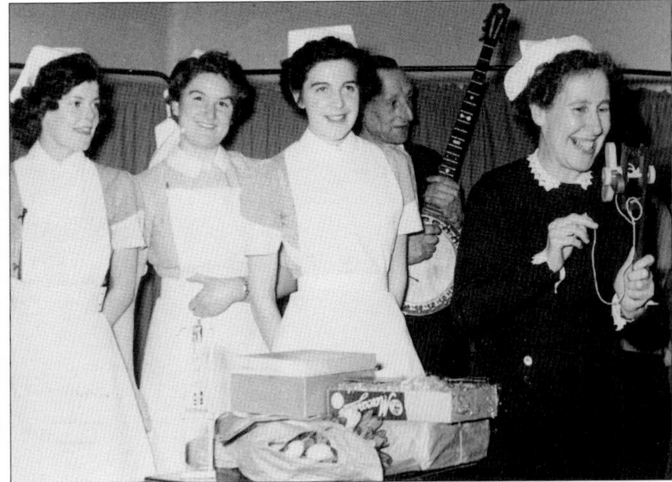

Luise as Assistant Matron in Dunfermline, Christmas 1960.

Wedding Bells. Left to right: Jimmy Harris (best man) Gerald, Margaret, Muriel Rodger (matron of honour), the Rev. McNeave.

A. Marion (aged twenty-one) in US forces, 1946.
B. Marion (aged sixty-seven) with Gerald in Baltimore, 1992.
C. Marion (aged twenty-two) with GI mates in Germany, 1947.

Gerald with Prof. and Mrs Ralph Goodell, Connecticut, 1992.

Left. Bon and Jai Nimbkar, India, 2000.

Below. Chanda Nimbkar with Gavan Bromilow, India, 2000.

Bottom. Gerald (in foreground) with itinerant sheep herders and their children, India, 1988.

Above. FAO consultants, Gerald and Mel George, with the yak project team at SW University for Nationalities, Chengdu, China, 1988 (Prof. Cai Li, extreme right).

Right. Gerald on back of a pack yak.

Below. Chinese Ministry of Agriculture delegation to the UK (Ma Ying in centre) on visit to Gerald and Margaret in Biggar, Scotland, 2007.

next day to the harbour at Hamburg. The whole place was alive with people trying to buy tickets for the next freighter to Shanghai. It was the only country left that Jews could emigrate to. But it was hopeless trying to get a ticket and Thea walked around for hours. Then she met a woman who was standing in the queue for tickets, and she was crying. When Thea spoke to her, the distressed woman told her that her husband had been very ill and had passed away two days before. She did not need the ticket for the boat leaving in two days and had come to sell it back.

Thea, in great excitement, implored the woman to sell the ticket to her. She told the woman about Paul, and how the authorities had promised to let him go if she could show them the ticket for him to leave Germany. To Thea's great relief, the woman immediately agreed to sell her the ticket. At this point, Thea broke off so that she could wipe her eyes, which were wet with tears, and compose herself.

She reminisced about old times in Breslau and how handsome her and Paul's father was, and about how happy the family had been. Her mother was so caring, and such a great cook, and the house was often full of friends and relations. She spoke of how they had a small department store and, at the side, there was a small cinema. They showed silent films, and Gerald's father would play the piano to accompany the action. Everything, she said, had been such fun in those days. Then she returned to that terrible effort she had to get Paul released.

Proudly Thea continued her story of how she managed to obtain some suitable cloth, and had sat up all night sewing new clothes for Paul so that he could leave as soon as possible. She sat back on her chair, a little breathless, her cheeks flushed. Vera patted her hand. Gerald and Margaret were shocked at this stunning story. Gerald caught the great emotion of his aunt Thea, and grasped her hand across the table and told her that she had been brave and good. He went on to say that he truly wished he had sent a different message in answer to the message he had received from his father through the Red Cross so long ago.

Thea insisted that Paul spoke of Gerald often and wondered what had happened to him. Paul, she said, had wanted to make a home for his first son in San Francisco. 'But,' she added, 'it wasn't to be.'

Gerald changed the subject by asking about their journey out of Germany to Shanghai. He would consider the details of Thea's emotional tale in private later.

Thea thought back on journey they had made in 1939. Because her husband was not Jewish, the couple were able to get a slightly better cabin on the ship. It also helped that she was quite pretty, and always open and friendly with everyone, and Vera was just a young girl. They still had all their belongings and some money. Although many of their possessions were stolen by the Nazis, some of their furniture and clothes were still there when they arrived in Shanghai.

It seems that the German crew on the boat were all right with the escapees from their country. One of them, who was particularly friendly with Thea's family, even told them which of the crew were Nazis so that they could be careful when around them. When they were still in Germany, Thea's husband, Fritz, was advised to divorce his wife or else get out of Germany. Gerald shook his head at this, hardly able to believe the story.

Vera's first experience of Shanghai was positive. There were lots of young people around her age. They lived in an international settlement, and they kept themselves apart from the Chinese. Vera felt sorry for the poor Chinese, who seemed to live on next to nothing and worked hard just to get by. She felt that because they had been well off in Germany things were initially not too bad in Shanghai. Her grandparents, fleeing from the Nazis, took the route by train through Siberia. They had a gruelling time, but eventually managed to join the family.

Listening to his cousin's story, Gerald thought of his own grandfather, and of his mother's sister, Erna, and her husband, Erich – Marion's mother and father – who had been left behind in Berlin, and in the last days of the war were murdered in Nazi concentration camps.

Vera explained that when they first arrived the Jewish people were not victimised, but when the Japanese bombed the American navy at Pearl Harbor, it became a different story. Aid from America ceased, and no more money was forthcoming from Jewish organisations. And the Germans who were already in Shanghai tried to pit the Japanese, who now ruled the roost in Shanghai, against the Jewish refugees.

After the Japanese invaded Shanghai, the Jewish people had to move from their comfortable houses into a ghetto in a much poorer area on the outskirts of the city. The housing there was very bad; there were no bathrooms, and no water in some places. There were bugs, and lots of horrors.

They were told that they had to have a pass to get out of the place to go to work. Everyone in the ghetto had to cross a bridge, which became notorious. There were guards on duty all the time. One of the Japanese was well known as an insane man. He screamed at people, and beat some up.

They had to sell everything to get money for food. And so they got poorer and poorer. Meanwhile, the Nazis were stepping up the pressure for the Jews in Shanghai to be handed over to the Germans. The Japanese military governor sent for the Jewish community leaders. The delegation included Rabbi Shimon Sholom Kalish. The Japanese governor was curious, and asked, 'Why do these Germans hate Jews so much?' Without hesitation, and knowing that the fate of the community hung on his answer, he told the translator in Yiddish, 'Tell him it is because we are oriental just like he is.' The governor, whose face had been stern throughout the meeting, broke into a smile. He liked this answer.

He did not accede to the German demand and the Shanghai Jews were never handed over. Besides, it seemed there was a history of Jewish help to Japanese people in past times. Also, they admired them because they got on with things in Shanghai. They opened little shops, and they had a synagogue, and a good school. Thea's husband had a tailoring business, and Paul, Gerald's father, played in cafés and nightclubs with his accordion band along with a few other musicians.

Thea wanted to emphasise that it was Paul who was the spirit of the family and it was he who pulled them through all those bad times. She added that, thanks to her own husband not being Jewish, Vera and she had passes that allowed them to move around freely. The less fortunate Jewish refugees, about 18,000 of them, had to stand in line, often throughout the night in the cold and wet, to try and get a permit to work. That crazy Japanese officer would give permits to those he liked and beat up those he didn't. And they all had to wear a badge with 'Jew' written on it, just like in Germany, but written in Chinese.

Later that afternoon, Thea told Gerald and Margaret the sad tale of Fritz, her deceased husband. It seemed that his health could not take the conditions in Shanghai, and he died in the heat of summer. It was such a sad time for them. There was no transport, no burial ground. Paul had to borrow a wooden barrow to take the corpse to a place where it could be buried. She heard a Japanese soldier ask, 'Who has died?' and the answer came, 'Oh, just another Jew!' And then Thea stopped, sobbing, 'But he wasn't a Jew. He wasn't a Jew!'

The irony and sadness of this story was hard to bear. In 1948 the family were finally able to leave Shanghai for the USA with support from a US Jewish Welfare organisation.

Emotionally drained with all these trying stories, the four of them – Thea, Vera, Gerald and Margaret – rose from the table in the restaurant and came out into the busy San Francisco street.

Gerald remembers how slowly and gravely they walked down the steep pavement. He knew that the pictures his two relatives had painted would stay in his head forever. On saying goodbye, Gerald thanked his aunt and cousin, and they embraced amid that crowded, noisy thoroughfare. These two had become dear to him, and he knew it would be hard leaving them.

As they were parting, Vera said that she would like them to meet a friend of hers who did not get out of Germany at the time of the Nazis and had had a terrible time. She had a personal story to tell of a kind that not many people had ever heard. She was meeting her friend in the Golden Gate Park the next day. Of course Gerald and Margaret wanted to be there and they arranged where to meet. Vera reminded Gerald to bring along his small dictation recorder that he had been using, as her friend's story was out of the ordinary.

Chapter 21

Revelations in Golden Gate Park

Gerald and Margaret soaked up the warm Californian sunshine and vibrant atmosphere around the Golden Gate Park. They laughed with the attendant who, talking through his little window, showed surprise that they asked for senior-citizen tickets. A sign with oriental style lettering said 'Japanese Tea Garden'.

At the Japanese tearoom, they met Vera and her friend, Eva (not her real name). They found a table, and with their tea they talked of the weather and the beauty of the planting in the wonderful park.

Eva was an attractive, smiling woman, who looked young for her age. She seemed more light-hearted than Vera, but this notion was dispelled as the afternoon wore on. Vera thought it best to leave Gerald and Margaret alone with Eva while she told of her wartime experiences.

Having been forewarned, Gerald asked Eva if she would mind if he switched on his small tape recorder so that he could listen again to her words when he got home. Eva agreed, on the understanding that if he were ever to recount her story, her real name and whereabouts would not be revealed. Of course, Gerald and Margaret agreed to this. When it came to the writing this book Eva gave permission to retell her experiences. The story came across like a fantasy and here it is in Eva's own words.

'I was a *mischling*,' she began. Seeing their puzzlement she went on, 'That is what we were called by the Nazis, because I was born to a Jewish father and a Catholic mother. Both my sister and I were raised Jewish. After Kristallnacht, when all the synagogues were destroyed and the shop windows of Jewish businesses were broken, my sister and I and many other Jewish children were made to change to Jewish schools.'

'That is what happened to me too,' said Gerald. 'But I was so young at the time, and it didn't bother me too much, as I remember.'

'Yes, you were lucky. Those of us left behind in 1941 were forced to discontinue our education altogether, and we became slave labourers. Yes, we worked for ten hours every day without ever having any days off. We were making uniforms and fur gloves for the German army. Then we were sent to a garbage dump to sort out the garbage. I got ill with appendicitis. I had my appendix out at the makeshift Jewish hospital. Our proper hospital had been taken over by the Nazis.'

Gerald stopped her and said softly, 'Drink your coffee, Eva. Take a break.'

'What I would give for a cigarette!' The tired woman sighed, holding her emotions in check.

At this Margaret took off to the kiosk and came back with a pack. The two females lit up the pungent American cigarettes as Gerald looked on silently.

'At first we were allowed to go home to our parents at night. Soon, however, we were picked up and transported to a slave labour camp. We heard that the Russians were coming, and when some of the local German people tried to run away, my sister and I took the opportunity to mix with the Germans and to try to escape.

'First we hid in a convent, but we had to leave there to save the nuns from being punished. The Gestapo were banging on the door when we left. We made our way to Breslau to try to see our parents. But the Gestapo were waiting for us there. We were transported to Buchenwald concentration camp, and then on to the Mauthausen one. They had given us civilian clothes, which surprised us. By this time we had met up with another girl, Susan, and the three of us were trying to stick together. On the journey from one camp to another, we were the only Germans on that train. All the rest of the passengers were Hungarian Jews.

'The train suddenly stopped. We heard bombs falling on both sides of us. We were at a place called Celle, close to Hanover. I was worried as my sister, Elsie, and our friend, Susan, were both sick. They wanted to try to escape from the train, but I thought they were too weak to run. I took a chance to talk to the guards and asked to see the commanding officer. When he came, I saw with surprise that he was not an SS Guard but a high-ranking German officer. I told him that the three of us were not Jewish, and that we had got on the train by accident when we were trying to escape from the Russians. I know that he knew that this was not true, but he said that as soon as the bombing stopped he would let us go.

'My sister did not believe him, and when the guards were not looking we got out of the train and started running.' She lit another cigarette and continued, 'I was a long way in front when I heard machine-gun fire. I turned round to see Susan and my sister lying on the ground. I shouted at the guards that we had been told that we could go by the commanding officer.'

There was a pause here while the trio took time for breathing space to cope with the story. After a little while, Eva carried on. 'They took us back to the train, and when the commander heard that we had gone on our own, he was very sad. He immediately called an ambulance. We were taken to the hospital in Celle. Susan died in surgery, and my sister died two days later.'

'I'm so sorry,' said Gerald. 'I should not have asked you to dig up all this sadness.'

'No. You are all right. I want to tell you. It is a true story, although only one of many.' After a pause, she went on. 'The Gestapo came to interrogate me, and I gave them the same story – that we were not Jewish, just on the wrong train. The doctors and nurses knew that this was not true but they did not give me away. A bit later the English occupied the town and I worked at the hospital and felt safe.

'Eventually, I heard that my parents were looking for me. I took off across the Russian-occupied sector, and close to the border I was stopped by the guards. I was gang-raped by them . . .'

There was a sharp intake of breath from the listeners at the shock of this revelation, but Eva carried on hesitantly. 'And then . . . they let me go.'

They took a break then. The park was quietening now, as the afternoon sun went down and mothers started taking their children home. Then Eva cleared her throat and started again. 'Quite soon after that, I met a guy who was also on the run. He said, "Look, we'd better get out of here fast." He had a motorbike and helped me get to the train station. There I got on a train and eventually got to a United Nations refugee camp. Most people were waiting to be transported to the United States.'

Her voice had lightened a bit now as the incredible story continued. 'There I met the man who became my husband. By a miracle he had survived the Auschwitz camp. We hit it off right away, and we clung to each other in those terrible days.'

Gerald reached out and put an arm around Eva. 'I am so glad you started finding some happiness. Your face is brightening now. You sure deserved some luck. What a survivor!'

'Yes, things started to get better then. My parents were in contact with my dad's brothers in America. They gave us affidavits; those were what all refugees craved at that time. They guaranteed to be there for us when we got to that country. And so we were allowed into the United States. We arrived on Ellis Island on 3 January 1947.'

Eva smiled now. Gerald was stunned to the core by this long tale. Most of the time he had been holding his breath in disbelief that such a normal, average lady could be telling such a story.

He managed to get some words out as he looked across the tea-room table. 'I am totally lost in admiration of you,' he stuttered. 'God, you were brave! And how sad that you lost your sister and your friend in that way. What an awful experience to go through.' Eva bowed her head for a few seconds as he went on. 'It must have been great to find your parents again.'

'Sure was! We laughed and cried for a week. But all those bad things, they're all in the past now. Water under the bridge, as they say, Gerald. Now I am in this beautiful part of the world, and I am very grateful to the USA for what it did for me and my family.'

Before they all parted that day, Gerald said, 'It's been great to meet you all and to hear your stories. Before this I had my mother and no one else for a family. But now I know I have some wonderful relations in the States – a real family and friends.' He had come to love them.

They hugged each other and Vera said that they wanted hear more about Gerald's experiences. Thea would particularly like to hear about Luise, as she was very fond of her. Gerald remembers promising to write to Thea when he got home and tell her all about his mother. They said goodbye at the gates of the park, smiling and with much hugging. Gerald said that he and Margaret would try to come back to this lovely part of California to see them all, and he hoped that they might in turn take a trip to Scotland some day.

Chapter 22

Luise's Story

This is an extract from the letter that Gerald wrote to Thea about Luise's departure from Germany and her arrival in the UK.

My mother had arranged her emigration from Berlin to England before I was put on the Kindertransport. She arrived in London in a small plane just before the outbreak of the war. She was very nervous and excited, especially as this was her first flight. I know she was anxious to see me, and find out if I was all right, but also she was very worried about my grandfather. She had looked after him for twelve years, and as he had moved to an old people's home before her departure from Berlin she wondered how he would cope with the change.

He'd been allowed to bring with him his favourite armchair, his clothes – he did not have many – and a framed photo of his darling, departed wife and of his daughters. But he could not take other mementoes. He had wanted to pack some of my grandmother's little ornaments, my teddy-bear, and the cushions so lovingly embroidered by Luise. None of the comforts of his lifelong surroundings was allowed. His other daughter, Erna, was still somewhere in Berlin but she was not able to give support to her father. She and her husband, Erich, had already been hounded out of their apartment and Erich had been taken for a time to a concentration camp, having escaped that fate once before, but now mercifully released. But they felt as if on the run, moving from one apartment to another, hoping the Gestapo with their random raids would not catch up with them again.

Later when I was older she told me of how great it felt to be out of Berlin with no yellow armbands and no fear of the uncouth Hitler youth taunting her in the street. My school friend's parents, Rudi and Herta Seidel, were waiting for her when she arrived in London. She was so relieved to see them, and grateful also, for they had been able to sponsor me for inclusion in the Kindertransport of unaccompanied children taken in by the UK earlier that year.

My mother told me how Herta flung her arms around her when she arrived at the airport in London. You must remember how tiny she was, only five feet with her heels on. They both cried a little and then Rudi, collecting Luise's one suitcase, took her by the arm, saying that she was to stay with them for a few days while Luise went to visit Horst and before starting her training as a midwife. My mother remembers him adding 'to bring yet more children screaming into this already crowded world' as Rudi always had a way of being funny.

Two days later, the train steamed its way to Margate and my mother found her way to Rowden Hall hostel where I was staying with other refugee children from the Kindertransport. Apparently she had to ask her way many times in her very broken English. She had bought the ticket for this journey already in Berlin. Money for a taxi was out of the question. She had the ten pounds she had been allowed out of Germany and another twenty that Rudi had given her in repayment for money loaned to them by her father years earlier in Berlin. This money would have to be treasured. She knew a trainee midwife would be paid only twenty pounds in a whole year – but she would have a room in the nurses' home and her food.

I ran out of the hostel as I saw my mother approach. There were lots of tears, mostly from my mother as I had adapted fast to my situation. It was a fine day in April and the beach was near deserted. We talked as we walked along the front. I will always remember how good was the taste of the ice-cream cone that she bought for me that day.

I know that she had a lot of worry and hard work, as she told me years later. The start of her two-year midwifery training was at the Radcliffe Infirmary in Oxford. It was hard for a woman in her mid-thirties to learn a new language, a new demanding profession with long working hours and to attend lectures and study books in order to pass the exams in both theory and practice. She had only ever kept house before and served in a shop.

Soon after the outbreak of war the maternity wing of the Radcliffe was cleared to make space for anticipated greater medical and surgical needs. Midwifery moved to a makeshift maternity hospital created in the requisitioned Ruskin College in Oxford, a short distance from the main hospital. I don't remember much from that time apart from the occasion when I, on a visit to Ruskin College, spotted Luise pushing a trolley with no less than twelve babies, some crying, from the nursery to their mothers waiting in the wards. I remember that comical sight with babies packed like parcels side by side, alternately head to foot, but my mother seemed to be enjoying it.

Part two of her midwifery training took Luise to Luton, a small town with none of the glamour of Oxford's magnificent college buildings and parks. But it led to a lasting friendship with a William and Shirley Sansom. He was the minister of the Congregational church in the town. He also conducted services in the hospital where Luise worked, services that she attended and where the friendship started. It was a curious fact of her life that she refused to renounce her Jewish religion, never again practising it yet finding comfort in Christian worship. She regularly attended the services in hospital throughout her career and occasionally also went to church. For her it was the same God of Christians as of Jews.

Her friendship of the childless Sansoms even extended to their having a close friendship with me as I grew up, and they treated me as a son. When they died they left me in their wills two of William's precious possession – a walnut bureau and a gold Longines watch – both of which had been presented to him by grateful congregations.

By this time Luise had applied for and been granted British citizenship, of which she was immensely proud. Her training completed, Luise became a staff midwife in a London Maternity Home with, for her, the princely pay of one hundred pounds a year.

Paddington Hospital, a large hospital in London, had survived the bombing, and it was there that Luise spent a further three hard-working years to train in General Nursing. Because of the war the minimum age for starting nurse's training had been raised, so Luise could fulfil this further ambition of hers. Fully qualified as nurse and midwife, she spent two years as a staff nurse in London's Hammersmith Hospital. There she helped supervise trainee nurses. Ambition yet again took hold of Luise as she applied for and became a ward sister in a large teaching hospital in Bolton, Lancashire. Her friends the Sansoms were by that time also there, as William had been called to a church in the town. I think that may have had something to do with her choice of Bolton.

Ward sister was not the end of Luise's drive to improve herself. She was not experienced enough to apply for a post as assistant matron or even matron but to get there she was determined. So the next best thing, equal in status but distinctly unpopular with nurses, was a post as night superintendent.

Thus she came to a sizeable hospital in fashionable Guildford, a charming smaller town in the south of England.

A year of working all night, sleeping all day, apart from the nights off – a gruelling regime for so long a stretch of time, but for Luise it had its reward. Two posts as assistant matron followed. One in a hospital for training nurses in Leamington Spa, the other post in Dunfermline in Scotland. From there she had the opportunity of seeing me and the family, as she was now in striking distance of Edinburgh.

I used to get angry if occasionally she slipped back into calling me 'Horst' or wanted to speak German. I would have none of it. And then, of course, there was my wife who she could get to know, and also be a grandmother to Andrew and Ruth, her two young grandchildren.

Many a Sunday she would invite us over for lunch. I remember taking our old car across on the ferry over the Forth. Sunday lunch was brought to her and any of her guests to her quarters in the hospital, a bedroom, a sitting room and her own bathroom. The food was always good. Amongst her duties with ward rounds, chatting to the patients, keeping nurses in order and keeping friendly with the doctors, she had overall charge of the domestic staff. She made sure of competent cooks and wholesome food for all the staff and patients. She was also a stern taskmaster, not allowing any lapses in the wards, but the nurses came to respect her for that as did the domestic staff who did not dare to fail in keeping the hospital clean.

She retired from the hospital in 1965. One thing she had always promised herself was that when she had the chance, she would go and visit Marion, the daughter of her sister Erna who had died in the gas chambers of Auschwitz. Marion had managed to emigrate from England to Baltimore in the USA. Sadly when Luise met up with Marion, then in her forties, she was in a poor state. She was in hospital being treated for schizophrenia. It seems that on hearing of the dreadful fate of her parents in Germany, Marion's health totally collapsed. My mother found that visit emotional and harrowing and on coming back home she never spoke much about it.

The word retirement was not part of Luise's vocabulary. She immediately set about taking on home-nursing and companionship jobs and, even during the tourist season in Edinburgh, working in a Princes Street shop where her knowledge of German was useful for many of the foreign tourists. She loved the shop work – it reminded her of her younger days. She had vowed to herself she would not use her retirement pension but would continue to work in order to pay for her needs and life in the council flat she rented in Edinburgh. She saved her pension and some more to achieve her lifelong ambition to buy and own a small flat of her own. This she did and was immensely proud of her achievement starting with twenty pounds a year pay in 1939 to a flat sold, after her death in 1990 aged nearly 87, for sixty thousand pounds.

She had always been a hard taskmaster to herself, not just to others. She spent money on nothing except the essentials for herself. But she did buy over-generous presents for her family. She never failed to be grateful for the new life she had been given in Britain and she took pride and joy in her friends and satisfaction in her career. Looking back, I would say that her thoughts and actions were mostly for her family.

The austerity that Luise practised on herself, her self-denial and her drive, was not attractive to some others. They found her standards too demanding. This was especially true of young people who could not, or did not, wish to emulate her drive and determination to succeed, or her willingness to forego relaxation and fun in the process. With the passage of time we have come to appreciate her legacy better. She died in an Edinburgh hospital in 1990 after a short illness.

Chapter 23

A Journey to Meet a Brother

Gerald had long wanted to see parts of the USA other than those he had visited whilst on his Kellogg Foundation fellowship twenty-five years earlier. He also was anxious to get to know his half-brother Jerry, who had been only a boy on that first visit to his family in San Francisco. He had not seen him since. Now Jerry was a man with a wife, Mari, and a teenage son, Darrin. They lived in Vista, near San Diego, and the journey there from San Francisco would allow Gerald and Margaret to see some beautiful parts of the country.

A day or two after their meeting in the Golden Gate Park in San Francisco, it was time for Gerald and Margaret to introduce some adventure to their travels. They left in their rental car for the famed Yosemite National Park. There are over seven square miles in the Yosemite valley, where they found themselves amidst spectacular scenery, the great granite cliffs, thundering waterfalls within touching distance, and the lovely clear streams and lakes to be found everywhere.

They learned of the wonderful work of the Scottish-American John Muir. They were told of his devotion to nature conservancy, and of how he became a friend of President Theodore Roosevelt. The president had read some of the writings of John Muir, and expressed a desire to meet the man. This meeting culminated in the two camping out together while Muir explained his theories of the geology of the region and of its importance to the nation. He convinced Roosevelt of the value of the preservation of the area. Because of the work and knowledge of this lifetime naturalist, the president passed a law that took the area under federal government control. Sheep grazing was stopped in the area and business exploitation was also prevented.

Next, Gerald and Margaret went on to the Kings Canyon National Park, and in Grant Grove were astounded by the giant Sequoia trees – among the oldest known to exist. A park ranger, immensely proud of his forest, instructed them and other tourists about the Sequoia tree's resistance to fire. He explained that the forests were cleared of unwanted undergrowth by fire while the Sequoias resisted the flames with their fireproof bark. That said, it is not unknown for fires to grow out of hand and burn for many months.

Gerald and Margaret continued their journey south towards Vista. On one of their short stops on the highway they were offered iced tea, free of charge, the shop owners intrigued by visitors from so far away. Many of the Americans that they met immediately claimed Scottish ancestry, and many had stories to tell of how they got to the States.

Gerald was anxious to press on and they had to cut short visits to many of the natural wonders that they would have enjoyed seeing. He kept his eyes on the road, eating up the miles in the growing heat of the summer. They talked of the Wiener family while they bowled along, including Gerald's two half-brothers. Both had been born in Shanghai: Pete in 1940 and Jerry in 1948, on the same day that he and his family boarded the ship for America.

They bypassed Los Angeles, stayed one night in Pasadena, and reached Vista at lunchtime next day. Jerry and Mari's house in Vista was beautiful. It was open plan, on one level, with a large billiard table at one end, a comfortable seating area with a television, a computer corner and very modern kitchen. Most of the back garden was taken up by a swimming pool. When the two couples got together, the conversation didn't flag. They talked all through the afternoon, and often the discussions became hilarious, as Jerry was a great one for practical jokes. He had prepared a scientific experiment to fox his academic brother. But of course it was a trick and, as intended, Gerald was suitably puzzled with the unnatural result, while, in the meantime, Mari and Jerry laughed uncontrollably.

The following day, Mari left for her teaching job at around seven in the morning. Jerry, also a teacher, had been able to take some time off and along with his son, Darrin, took them on a tour of San Diego. From a viewing point they saw the whole stunning bay.

At eight the next morning, Jerry, Mari, Darrin, Gerald and Margaret stood at the gates of the first of the Disneyland parks ever opened, on the outskirts of Los Angeles. The Scots visitors were full of excitement, whereas the Yanks had been there a few times before. The group took a tour of the complex and were astounded at the sights – simulated space travel, underwater tours in a submarine, cruises on an ocean attacked by pirates. And, last of all, at ten o'clock, the exhausted party stood in amazement as all the Disney characters paraded in the darkness of the evening.

On the drive home, while Darrin and Margaret dozed off, Gerald and Jerry got to talking about their father. Jerry said he remembered him being a lot of fun throughout his childhood, and Gerald envied his brother for an opportunity that he did not get. Jerry thought that Pete would be able to tell Gerald more. What he seemed to recall most clearly were visits to Chinese restaurants with the family. Gerald supposed that was why Jerry loved Chinese food. Jerry agreed and went on to ask Gerald if he thought it was funny that their father had called them both by almost the same name. They puzzled over this. Gerald's father had been described to him as an extrovert, a sociable, fun character, and he temporarily felt jealous of his brothers. They arrived back in Vista in the small hours, all falling into bed and into a deep sleep of exhaustion.

It was planned that, the following day, Gerald would exchange his hired car for a better one. Gerald told the rental clerk that the first car was a bit boring, falling short of the advertisement, and that he would be crossing the desert as far as Denver. He hoped for something more exciting. The rental clerk recommended a white Cadillac, telling Gerald that it had cruise control and temperature control, and that it was the tops. The best part came when the clerk said they would charge no more for this hire than for the earlier one, to make up for the disappointment Gerald had expressed.

The car was beautiful and Gerald was immediately thrilled at the idea of driving it across the country. But it was of greater length than any car he had owned before, which made him apprehensive. It was also an automatic, like most American cars, and he was not very familiar with this style of driving. That day he drove everyone very carefully to San Diego Zoo, and told Jerry that he did not think he could keep the car for the long trip. Jerry replied that he would soon get used to it, and it would be great for a trip across the desert.

Chapter 24

A Sentimental Destination

They left Vista the next day in the white Cadillac on what was to be a wonderful adventure. Feeling like millionaires in this beautifully cool car the pair arrived at Laughlin, Nevada, a small desert gambling town recommended as the poor man's Las Vegas and a convenient stop on their way to the Grand Canyon. They parked the Cadillac and hurried through searing heat into the ultra-cooled lobby of the Flamingo Hilton.

They passed through a sea of gambling machines, brightly lit and noisy. In front of each machine sat someone mesmerised totally by the revolving images. The reception desk stretched all the way down one side of this huge lobby area. The room they were allocated seemed enormous, with a mirrored ceiling above the four-poster bed. This caused them much hilarity.

The strange thing was that in spite of the luxury on every side, and the very good food, the cost of staying there was very low. Obviously the hotel made its money from the casino, but they lost out with Gerald and Margaret. When they checked out at eight o'clock the following morning, the one-armed bandits in the reception lobby were still clattering on, and gamblers never stopped staring at the whizzing pictures of bells and lemons, hoping for three in a line and for the cash jackpot to come streaming out. It was all a bit creepy, and the pair were glad to get back into the morning sunshine and onto their planned route.

This took them next to the Grand Canyon. They had read and heard about its grandeur and they were not disappointed, but were overawed by the huge chasm before them. As Gerald refused to take a guided tour into the canyon or the helicopter flight through it, the four days they had booked for their stay felt a bit too long. In the end, magnificent as the views were from a multitude of tourist viewpoints, Gerald found that there was a limit to the number of photographs he could usefully take.

At their next stop, in Flagstaff, Arizona, they spent ages trying to find the actual flagstaff, which had been raised by the settlers of the past to mark the place of a spring, or so their guidebook told them. None of the locals seemed to know. They had to go the local government offices before they were directed to this historic little memorial.

In the desert they stopped at a Native American settlement where a trading post was selling locally crafted jewellery. Necklaces, bracelets and earrings of beautifully carved silver inset with turquoise precious stones were on sale, and Gerald bought a set for Margaret as a memento of their trip across the desert.

One night was spent in Albuquerque, New Mexico, and part of the day exploring the lovely town of Santa Fe. There the Spanish atmosphere was seductive. Many of the residents were retired people who had sought out the peace and beauty of the place.

They arrived in Colorado, and having enjoyed a piano bar evening in Santa Fe, enquired in their new hotel, some distance outside Denver, if there was any bar playing music in the area. The next few hours were memorable for them. The bar they were directed to proved to be a great surprise. The place was called 'The Golden Bee, An English Pub'. The actual bar was a very long, curved construction of solid oak. It had been imported from an old London pub many years before. This was hard to envisage, but apparently it had been reconstructed to appear exactly as it had in London. They learned that in the nineteenth century, Colorado was a place where younger sons of landed gentry from England and Scotland were sent to try to make their fortunes. There was therefore a strong influence from the United Kingdom in parts of the state.

In front of the bar were round tables for four. Here meals were served, and at half past eight in the evening songbooks were given out and a pianist started to play. To the astonishment of Gerald and Margaret, the pianist began with the 'The Bonnie Banks of Loch Lomond'. The patrons joined in the singing, and the atmosphere was cheerful all round. As they were leaving the bar, they stopped at the piano, and said, 'Thank you for playing so many Scottish songs.' The pianist was thunderstruck. 'You're Scottish? I don't believe it. My grandfather came from Scotland.' Another man called out, 'My uncle is Scottish. He supports Rangers football team.' The pianist, to their amazement, began to play 'God Save the Queen', and then he insisted that they accompany him by singing 'The Skye Boat Song'. Gerald was more than happy to let Margaret do the singing. They left amid applause and cheers. It turned out to be one of the most fun evenings of the trip.

Gerald wanted to make a sentimental return journey to Madison, Wisconsin, where he had spent those happy months on a postdoctoral fellowship. He had been back once for an international conference on trace elements but Margaret had never been there. Gerald wanted her to get at least a glimpse of the attractive town and the delightful university campus beside Lake Mendota.

Art Pope, the 'sheep man' at the university in those early days, had remained friends with Gerald and had said he was welcome 'at any time'. He and his wife, Betty, were at the airport to meet them and took their visitors to the small farm at Verona where they now lived. Art was retired now from his role at the university but his passion for sheep was undiminished. This was unusual for the area in which he lived, as this was the 'Dairy State', where every farmer had cows. However, Art had a flock of sheep of which he was very proud. It happened that the visit coincided with a 'sheep day' being held for farmers north of Madison and of course Art and Betty wanted to take their visitors to experience this folksy event. Gerald was bounced by Art into giving a brief talk to the farmers on sheep in Scotland.

After wandering round the exhibits and stalls sampling lamb burgers and, more to their taste, the famed Wisconsin ice cream, they headed back to Art's farm.

There they spent two more days with their hosts and on one of them they went to Madison, a pilgrimage for Gerald and a new experience for Margaret. As hoped and intended, she was impressed by both the town and the charming university campus.

Sadly for Gerald, Professor Chapman, his mentor during his study trip, and his wife were away on holiday and he could not introduce Margaret to that lovely couple.

They left Wisconsin to fly to Baltimore to make good a promise Gerald had made to himself and to his mother.

Chapter 25

Marion's Story

When they arrived in Baltimore, Gerald hired another car – not as grand a one as the Cadillac – and drove to the Concorde Apartments. It was here that Marion, Gerald's cousin and childhood playmate from the old days in Berlin, was being looked after.

Both being only children, Gerald and Marion had felt more like brother and sister than cousins. Before leaving the United Kingdom, Gerald had shown Margaret photographs of Marion taken when she was a twelve-year-old girl. In one picture she stood, dressed in a velvet coat, with her prosperous-looking mother, both staring seriously into the camera. Now, they were moved to see what had become of her. Sadly, the good-looking, happy child had grown up to be an overweight, slightly confused woman, perhaps wondering why she spent her life in this little room alone. Indeed, when Gerald thought about it, Marion's was a tragic story.

Marion's parents had been scared by the Nazis' hounding of Erich, Marion's father, and by 1938 the family had obtained an affidavit allowing them to apply for a visa to emigrate to the USA. Furniture and belongings were packed and dispatched, and a couple of heavy gold bracelets sewn into the upholstery of one of the settees – these would be their only valuable items when they arrived in the States. But that day never came for them. Entry to the USA was strictly regulated by a quota system and their number had not come up by the time war broke out in September 1939.

Marion, however, was more fortunate than her parents. She was allowed into the UK in July 1939, on the strength of the fact that she would only be 'in transit' to the USA. She was fifteen years old and could take up work to keep herself. She arrived in Swindon in the south-west of England and began an apprenticeship as a hairdresser and beautician. She loved that and thought it could be her livelihood when she reached the USA.

In England, Gerald and Luise did not have much opportunity to see Marion. At the end of the war, she visited them when they happened to be in Oxford on a brief holiday at the Spooners'. She sprang the news that she would be leaving for Baltimore in the next month to stay with her father's cousins. She was excited because she had enrolled in the American forces and would be joining her unit in Germany soon after that. She hoped she would be able to find her parents when she got to Germany.

And so Marion joined the armed forces, went to Germany and for a while revelled in her new role. She sent Gerald a photograph with her GI mates, standing in front of a former Luftwaffe bomber that bore the swastika emblem. Soon she was to trace her parents, but not to where she had hoped. It was in Auschwitz, the notorious extermination camp, that she found their records. Erna and Erich went to the gas chambers exactly one week before the camp was liberated by the Russian army. The Nazis were good at recording everything, even their atrocities.

Marion spent two years with the forces, was demobilised, and returned to her father's cousins in Baltimore a much-sobered person. The trauma of her parent's untimely and cruel death had a devastating effect on Marion, triggering schizophrenia. The Jewish families organisation in Baltimore took over responsibility for Marion, partly funded by restitution money from Germany that their legal team had managed to extract on Marion's behalf. Marion remembered about the gold bracelets that had been hidden in the furniture, but sadly her father's cousins, tired of paying for the storage, had sold the furniture. They protested that they should have been told about the gold bracelets.

Marion spent the rest of her life either in a psychiatric hospital or a care home. During lucid periods she would correspond with Luise, her closest living relative, and with Gerald. The Jewish care worker explained that when they had asked Marion what she would wish to happen when her cousin and his wife came to visit, she said that she would like them to take her to a good fish restaurant. That is what she often dreamed of.

This was a happy yet sad occasion. The restaurant where they ate was relatively empty, and strangely cold, almost clinical. It seemed to be a fish shop as well as a restaurant. The white-uniformed assistants were constantly cleaning the counters and tables. Whether this was a Baltimore-style place or some sort of German-style restaurant left them puzzled. However, they quietly went along with the outing for Marion's sake. In the uncomfortable environment of the restaurant, the conversation was stilted. It was difficult to know just what to talk about. In a sombre mood, they drove back to the building where Marion lived.

They returned the next day thinking Marion might like to do a little shopping, but she had few needs and her only purchase was a box of chocolates, for like her cousin she had always had a sweet tooth. Gerald had the bright idea that a visit to Baltimore zoo might bring back happy memories of the times he and Marion had been to the zoo in Berlin. The visit was not a very great success as the sparkle had left Marion and even Gerald could not rekindle the excitement of their young days. They returned to the Concorde apartments and, with some emotion, said a tearful goodbye.

Two years later the letter arrived in Scotland from the Jewish care worker to say that Marion had passed away. She was indeed a casualty of the war, though not in any book of such records of the time.

Chapter 26

Catching Up With Old Friends

Gerald had become adept at driving in the USA on the right-hand side of the road. It was made easier because the roads were wide and well policed to prevent speeding. They left Baltimore and headed for Silver Spring in Maryland.

Being apprehensive about meeting more friends, scientists in particular, Margaret protested that Gerald knew too many clever people. Gerald had to admit that this was true. Clair Terrill, whom they were about to visit, was one of a small band of distinguished animal geneticists. He had graduated from Iowa State University in Ames, and in his own field he now ranked alongside some of the other greats from that university, including the doyen of animal breeding plans, Professor J. L. Lush.

Clair had overseen the vast US Sheep Experiment Station at Dubois, Idaho, where Gerald had first met him in 1957. The sheep there were occasionally at the mercy of grizzly bears and often of coyotes. Clair hated coyotes. Later he became the national programme leader for sheep and other animals for the United States Department of Agriculture.

They had a great welcome from the Terrills, Clair and Zola, and enjoyed an evening of celebration and reminiscing. The next day, Clair took them to Washington, D.C. They spent some time viewing the impressive Capitol Building and the beautiful White House. Clair then showed them around the National Museum of Natural

History, a place of great interest to the travellers. They lunched in the exclusive members' dining room. The whole day was a great treat. One of Clair's most interesting comments, coming as it did from a man concerned with large livestock, was that the success of poultry breeding in recent years and the cheap cost of chicken were what had kept many millions of poorer people in the world from hunger.

New York, their next stop, was a whirlwind in which they tried to take in as many of the famous sights of the city as possible. First of these was the great Gothic building which is St. Patrick's Cathedral.

Gerald knew that Margaret would be excited by this visit. They sat in one of the pews while Margaret told him of how her father's sister had moved to New York to work as a nanny for millionaires in the early thirties. The family she worked for paid for her wedding outfit so that she could have a beautiful wedding in this lovely cathedral. Thomas, her fiancé, had come out especially from Scotland to marry her in case she fell for the American life and married someone else.

They enjoyed Central Park with the joggers, cyclists and roller skaters. That day, by chance, they saw a long parade of German New Yorkers, dressed in the costumes of various regions of Germany. It seemed to be an annual German day. They crossed Fifth Avenue to visit the Metropolitan Museum of Art, then went to the top of the Empire State Building and then took in the Trump Tower where, from the atrium, they could see a waterfall flowing down the pink, white-veined marble wall from five levels above.

Gerald and Margaret wanted to see a Broadway show and managed to buy the last two available tickets for 'Crazy for You', a new musical then, which used the music of George Gershwin. This was a wonderful, thrilling occasion for them in the vintage Shubert Theatre.

Another unforgettable occasion was the evening they dined at the Smith & Wollensky steakhouse. Their eyes popped when, at the table next to them, four young diners had large white napkins tied in front of them by the waiters. Then four steaming, pink lobsters were brought to their table. More eye popping happened for both of them when their own prime steaks were brought. Each plate held a steak about two inches thick and weighing at least one and a half pounds. To them it was like a Sunday roast on each plate and they managed to eat only a fraction. Nevertheless, they ordered desserts: Margaret a small sherbet dish, but Gerald a sizeable apple strudel. He was suitably pained and ill that evening until he had walked off the effects of his overeating.

They had to get to Connecticut the next day to see two other friends of Gerald's from former times in Edinburgh. They managed to get themselves and their luggage on to the bus, and in no time they were saying farewell to the excitement of New York. They travelled north, through the Bronx and on through upstate New York. Soon Margaret started to enquire about the kind of people she would be spending time with during the following few days.

Gerald told her that Ralph Goodell was about the same age as himself, that he was a university professor and that he had met him through the church. Dinah, his wife, was very nice and sociable. Although she had attended Vassar College, which was known to be very posh and upper class, she was not like that. He remembered her as being a down-to-earth lady. Gerald had known Dinah and Ralph very well over the years that they had been friends in Edinburgh, where Ralph had been doing a PhD. He told Margaret that she would be amazed by the eighteenth-century house that Ralph and Dinah lived in. He knew all about it from Ralph's letters.

When the bus drew up at Willimantic, Connecticut, a tall, rangy man with a face like George Washington awaited them. He was eager to meet his old friend, Gerald, and to be introduced to his wife. The pleasure of the meeting was briefly marred as one of their bags had been stolen from the bus at one of its intermediate stops.

Ralph and Dinah were a kind and educated, if unusual, couple. Their house, situated in a clearing in the forest, had been bought by Ralph, who had a passion for old New England houses. He had dismantled the wooden building from the small rural town of Barkhamsted and, with help from his family, had reassembled it in Coventry, Connecticut. It took them some years to complete. Modern plumbing was installed, and electricity. But the outside appearance was as it had been in the eighteenth century. To step into this house was like stepping back in time.

There were wooden floors and uneven white plaster walls, and the doors made from planks of wood to which had been added old-style metal latches. They had to mind their heads when passing through the low doorways. There was a steep, narrow stair to the upstairs bedrooms, and to one of these the two travellers hoisted their luggage.

Dinah was interested in alternative medicine, as was Margaret, and Ralph had a Christian outlook that chimed with Gerald. He had a caring attitude, particularly towards countries like Afghanistan, where he had taught English. He had also fought in the Second World War, and in the famous Battle of the Bulge.

The next day in New England was unforgettable. The four friends started their day by exploring the garden and the surroundings of the beautiful woodland where the house stood. From there they drove to Sturbridge to be treated to another step back in time. Ralph had thought this would appeal to Gerald especially with his agricultural interests, but in fact it was fascinating for both his guests. Here was recreated an eighteenth-century village, complete with church and parsonage, and of course a schoolhouse. Visitors could chat to the costumed inhabitants of this living museum – the coppersmith, the blacksmith or the man who made barrels.

They wandered in and out of the houses and the village shop. At the gardens, they were advised about growing vegetables and shown the technique of preserving produce from the garden. The staff were anxious to impart all sides of living at that time in America. In the restaurant, children and their parents were shown how to cook in an old-fashioned way, and encouraged to taste the lovely food that was produced.

When there was talk of a visit to the Mark Twain Museum, Gerald and Margaret were immediately keen. Margaret had quite recently read *Adventures of Huckleberry Finn* for an English literature course and was a great admirer of the American writer. Gerald had read the book as a boy just as an adventure story, without realising that it rated as great literature. This was another great day out with the Goodells, which ended with a trip to the museum built in honour of the great twentieth-century artist, Norman Rockwell. He was well remembered by Margaret from illustrations that she had admired in American magazines in the 1940s. These colourful illustrations, throwing a light onto the luxurious American lifestyle, were an eye-opener to the British and European generation growing up after the war. Ralph was keen that their visitors should also savour some of the American history enshrined in their part of the United States. So the last two days of this wonderful visit were to be devoted to that.

After one of Dinah's healthy breakfasts, starting with millet porridge, they set off for Newport, Rhode Island, a charming seaside town dating back to colonial times and still with many colonial-style houses on its streets. Trinity Church, with its lovely wooden spire gleaming white, and in the style of some Christopher Wren churches in London, had to be seen. The box seats and central pulpit were indeed of a bygone age. The queen, they were told, had once worshipped here. The visitors were impressed.

Sadly the last day of Gerald and Margaret's stay arrived. They were taken to Concord, Massachusetts, where their first stop was the replica wooden 'North Bridge'. Among the first shots of the American Revolution were fired there against British regular troops who had come to the town to retake arms that they claimed were stored in the town. The British retreated. Now the town retained much of its historic charm with interesting buildings flanking a wide main street.

Ralph insisted that they could not leave the USA without seeing Boston, which was to be the last stop of what had become an epic journey. On the way there in Ralph's car they were given a glimpse of the Harvard University campus, the oldest and probably most prestigious of the American private universities. The buildings were clearly designed to impress, but Gerald thought them without the charm and elegance of the Oxford colleges he had come to love on many a visit with the Spooners. Of course he said no such thing to his hosts.

Their time in Boston itself was brief. Gerald and Margaret wanted to see, at least from the outside, the bar made famous by the television series *Cheers* – though this was not particularly the scene for the Goodells, who good-naturedly drove them past. Then they went for a final and splendid view of the city and the ocean beyond from the viewing platform on the fiftieth floor of the Prudential Tower.

The moment of parting had arrived. It was a happy parting after several joyful and interesting days for the visitors, but sad also as none of them knew if they would see each other again. Ralph and Dinah set off back to their old farmhouse and Gerald and Margaret to the airport for the first leg of their flight home to Scotland.

Chapter 27

From Glory Through Turmoil to Dolly

Gerald's appointment, in 1947, to the National Animal Breeding Research Organisation, as it was then called, came within weeks of that organisation setting up its headquarters in Edinburgh, though he chose not to take up his post for another three months, as described earlier. He was to play a not-insignificant part in its life for the next thirty-nine years. It will be useful, however, to describe first why the organisation was created and why it was set up in Edinburgh.

During the Second World War it had become apparent to the government that the United Kingdom, as an island, was too reliant on imported food for its survival. There were many short-term measures to increase food production at home, such as the vast increase in allotment holdings (often on previous parkland or waste ground) where many town-dwellers could grow their own potatoes and vegetables. The government slogan was 'Dig for Victory'. In the countryside there was the drive towards more arable production from previous pastureland, as well as the widespread recycling of food waste to be turned into pig feed. War Agricultural Executive Committees were set up in each county to aid this process. But a long-term strategy was also needed. A group of seven wise men was set up jointly by the Agricultural Research Council (ARC) and the Agricultural Improvement Council in 1943 to consider the needs for research in animal genetics which might in turn lead to improved livestock production.

Thus was born a new organisation under the auspices of the Agricultural Research Council. In 1944 the Council appointed Professor R. G. White as the director and C. H. Waddington as chief geneticist of the new organisation.

White accepted on condition that he could visit relevant departments and persons in the USA to assist in formulating a potential programme for his organisation and he did not take up his appointment as director until 1945. He was a tall, kindly and distinguished-looking man who exiled himself only reluctantly from his beloved Wales to take on the job. He was also delightfully still of the old school that addressed everyone by their surname – without prefix of Mr or Miss or whatever – at a time when calling everyone by their first name was becoming common.

C. H. Waddington, a one-time lecturer in Cambridge with a reputation both in genetics and as a thinker outside the box, and now chief geneticist of the new organisation, set about creating around him a group of research scientists and geneticists who could undertake theoretical and laboratory-based research in the new organisation, and they moved to Edinburgh. Waddington had also worked for the military during the war, in operational research. Many of the team of geneticists and biologists who came to join him in London had also been backroom boys on operational research during the war.

But how the new Research Organisation came to establish its headquarters in Edinburgh deserves a moment's explanation.

Edinburgh already had a reputation for animal genetics research derived from the Institute of Animal Genetics at Edinburgh University, founded in 1928 under the direction of Professor Crewe, but with antecedents dating back to 1911. Then, in a smart move by the university, Waddington was appointed to the then-vacant chair of animal genetics in Edinburgh. This ensured that Waddington moved his team to Edinburgh, rather than to a rival university. That also helped to attract the new headquarters of the National Animal Breeding and Genetics Research Organisation to Edinburgh. Soon thereafter the word 'National' was dropped from the title, to create ABGRO, to prevent the impression being formed that this might be a purely Scottish organisation, as it was a UK-wide one, albeit with headquarters in Scotland.

Waddington's newly assembled team was located in the University's Institute of Animal Genetics alongside the distinguished staff already there. They were then joined, though at a different location in Edinburgh, by the group, including Gerald, that was being appointed for studies with farm animals. This group was housed in a large stone-built villa, Glenbourne, in South Oswald Road. Over time an annexe of prefabricated huts was added behind the villa, with rooms and laboratories for an increasing number of scientific and technical staff.

The secretary and treasurer of the organisation, Mr James, had previously been secretary of one of the War Agricultural Executive Committees. Though he undoubtedly did his job to a high standard, there was always a feeling that he never felt quite reconciled to the free and easy ways of the science staff; in his previous job he had always called the shots. Even though the research staff worked hard and often long 'after hours', they would, alarmingly in his opinion, set off during 'working hours' without so much as a 'by your leave', going to lectures or seminars, or to take a visitor to see the sights of Edinburgh, or worse still to attend a furniture auction for a newly set-up house – there were several recently married couples on the staff. Fortunately the director and Hugh Donald, the head of research, were tolerant of such behaviour. They judged their staff on performance and not on time spent sitting at a desk. In much later years the overlord administrators in London decided that all newly appointed staff would need to clock in and clock out, just like in a factory. It is doubtful whether it improved research initiative and output, but the civil service bosses were satisfied. Gerald, as one of the old guard, was never subjected to this new routine.

On the whole these were matey times at Glenbourne. Private cars were still uncommon and whenever anyone acquired one, usually some ancient model, people would troop out to inspect the vehicle and congratulate the owner. Only Mr James had a posh car – a sporty, red Alvis.

Gerald recalls a day, when chatting with him, that he discovered the man had a secret dream. This was to retire to a country cottage in Devon with a beautiful garden of roses, which he would name 'Rose Cottage'. And he did just that. Gerald thus saw a softer side of this lifelong bachelor than had been seen of him as ABRO's secretary and guardian of the finances. Unhappily he died just a year after retiring. Mr James was succeeded at ABRO by Alan Totty, a very different kind of person – happily married, intellectual, and eminently approachable. In later years, when Gerald had been given periodic administrative responsibilities, he found it a pleasure to work with Alan and receive his sage advice.

There is one bizarre tale from Glenbourne that remained fixed in Gerald's mind: conditions were somewhat cramped in the annexe and one of his colleagues, a veterinarian by training, must have found lack of space particularly irksome. He was interested in the reproduction of pigs and at that stage he wanted to count the number of placentas in the uteri of sows, as related to litter size. To that end he collected, from the local abattoir, the uteri of sows after they had been slaughtered. Without a proper place to carry out dissections he used to nail the uteri to the door post of his room, on the corridor side, and proceed with his dissection from there. His room was opposite Gerald's.

The stage was set for Edinburgh to flourish as a world-class centre in animal breeding and genetics. This was further helped by the formation, in 1947, of the Edinburgh Centre for Rural Economy, with an estate of 3,000 acres and an interest in animal breeding among its many concerns, spanning teaching, advisory and research. This was the Bush Estate, situated a few miles south of thecity.

Professor White retired as director of ABGRO in 1951 and was succeeded by the joint appointment of Hugh Donald and Professor Waddington, each responsible for their own sections. Shortly thereafter Waddington's ARC-funded group was reconstituted as a separate Unit of Animal Genetics, independently administered but still financed by the ARC. And so, the word 'genetics' was dropped from the title and ABGRO became ABRO with Donald as director. Donald was given permission to have a new headquarters built for ABRO at the university's King's Buildings site – diagonally opposite to the Institute of Animal Genetics. The premises at Glenbourne had outgrown their usefulness.

The new headquarters was designed, at the instigation of Donald, by what was then the most prestigious firm of architects in the UK, Basil Spence & Partners. They were responsible, among others, for the post-war reconstruction of Coventry Cathedral, which achieved worldwide acclaim for its design. The two architects from the firm who were most closely involved with ABRO were John Hardie Glover and William Hunter. The building was notable for the black steel girders on its exterior, which supported the offices, laboratories and lecture rooms. The building, which opened in 1964, was entered through a large and imposing high-ceilinged hall with glass from floor to ceiling at front and back.

The Institute of Animal Genetics, in partnership with ABRO, attracted top students, and in its turn sent out, to countries like Australia and New Zealand and those in the Middle East, Africa and Asia, well-taught and enthusiastic young academics. It also attracted research workers and academics, many of them world famous in their discipline, for periods of sabbatical study and lecturing, many from the USA.

For much of its existence, a mainstay of research in ABRO was the projects on cattle, sheep and pigs, designed by Professor White and H. P. Donald. These experiments were located at the six farms that had been purchased across Scotland, England and Wales: two hill farms – Stanhope in the Scottish Borders and Rhydyglafes in North Wales; the large semi-upland farm of Blythbank some twenty-five miles south-west of Edinburgh, with a satellite upland farm, Broughton Knowe, not far distant; a dairy farm, Cold Norton, in Staffordshire; and two pig stations – Mountmarle at Dryden in the village of Roslin and Skedbush in East Lothian. In addition to that was the field station, also at Dryden, which housed large animal facilities for intensive studies. Dryden was located in the grounds of the university's Bush Estate.

Among the first senior staff appointments were what were termed 'husbandry' officers. They had scientific training but also wide experience with cattle, sheep or pigs (one officer for each species). They were to help the director and Hugh Donald in setting up the farms and later to act as a liaison between the farm managers and staff and the research scientists who were not credited with having practical experience of animals. True for some, but not all. The husbandry officer most closely related to Gerald's later sheep research was Joe Read, universally liked and known, even by the farm managers, who might have resented any interference, as a delight to work with.

The first tranche of scientific staff was appointed specifically to investigate and develop different aspects of the long-term farm experiments. Several had been post-graduate students at the Institute of Animal Genetics and many became leaders in their field. To develop the field laboratory, Hugh Donald had invited an Australian wool scientist, H. B. Carter, to take charge. Initially it was work related to wool and to resistance of sheep to cold and heat (in climate chambers) that predominated. But Carter and Donald, of equal seniority, never quite hit it off. After Carter left, the research at the laboratory broadened out into other areas. New laboratory-based enterprises were established in ABRO including a cattle blood-typing service for the industry.

In a sense, ABRO was a benign autocracy where, with only one or two exceptions, all the largest of the animal breeding experiments had been initiated from the top. Gerald's long-term sheep experiment was the first of these exceptions. But Donald was the final arbiter of what went on – all the scientific staff were directly responsible to him. That changed after Donald retired as director in 1974 and was succeeded by John King.

John had been the lead scientist in the pig research, and, as the new director, decided to set up departments within ABRO. One of these was to be the Department of Physiological Genetics. It was the largest of the departments created and Gerald was made its head. Its work was primarily concerned with issues underlying genetic variation in reproduction, production and survival traits. John's own passion, and the motivation behind his research outlook, was the application of genetic knowledge to the improvement of livestock at farm level.

Each publicly financed research institute was subject to an inspection every five years by a peer group of academics, advisors and farmers – euphemistically called a 'visiting group'. ABRO was subject to one such 'visit' in 1980. The group's report raised disquiet at the Agricultural Research Council about the balance of the research programme at ABRO. The more applied end of livestock improvement was also the remit of other organisations in the country, such as the advisory services with husbandry farms at their disposal, as well as some universities and colleges of agriculture. ABRO, it was felt, should be in the best position to undertake the more fundamental end of the animal breeding research, which might otherwise be neglected. This in turn led the ARC to lose confidence in the leadership provided by John King. John had become victim of a change of attitudes towards research funding rather than to any lack of personal ability.

Early in 1982 the ARC arranged a meeting with senior staff at ABRO, chaired by Professor Rook, deputy secretary of the Agricultural Research Council, along with a Mr Malcolm Carpenter (from ARC headquarters). Immediately following that meeting Gerald was requested by the ARC secretariat to write a personal and confidential assessment to the secretary of the ARC, Dr Riley, and to the past chairman of the visiting group, Professor Jinks. Not long thereafter Professor Rook revisited ABRO in the company of Mr Alexander, the permanent secretary of the ARC. They spent approximately two hours in discussion with Gerald and asked, amongst other things, whether he would be willing to take on the directorship of ABRO.

It needs to be mentioned that the ARC had previously indicated that they wished to reduce the size of ABRO both in terms of staff and facilities to approximately one third of its strength, to restructure the research programme along more fundamental lines, and to sell the farms. At that time the total staff of ABRO numbered 176, of whom sixty-one were in scientific grades. Over twenty of these were the lead research scientists. The remaining members of staff were in important support roles and around fifty of them were on the farms. The intention to shrink ABRO to but a shell of its former self led to a furore among the staff. A campaign was mounted aimed at industry leadership, politicians and the agricultural press, fronted by John King and Charlie Smith, to rescue ABRO from that fate.

Gerald had obtained a commitment from John Rook and Mr Alexander that the staffing would not be reduced below sixty per cent of current strength and that two farms – Blythbank and Mountmarle Pig station – would be retained as well as the Dryden Field Laboratory in Roslin. For reasons never fully clear to him, the ARC also wished to build a new headquarters and laboratories for ABRO at the Dryden site and abandon the King's Buildings site.

These facts became known at the time to Roger Land and Alan

Totty. The offer to Gerald was also known subsequently to his close colleague John Woolliams (later Professor) and of course to members of his own family. After a period of reflection, Gerald declined the offer of the directorship for two principal reasons. Firstly, the external 'campaign' to 'rescue' ABRO was very vocal and seemed to have wide support among staff. Gerald said that he would have felt a pariah to stand against that, in spite of likely support for change from among most of the senior staff. Secondly, he did not feel temperamentally suited to making a large number of staff redundant – even though that number, through Gerald's persuasive arguments, was far fewer than originally intended by the ARC. An added reason was that he would have felt (and been seen as) little more than a caretaker with only three years until he reached the statutory retirement at age sixty.

In retrospect, he regretted his decision, as he felt he might have contributed positively to the transition of ABRO into a new era. No doubt it would also have given him added status, including an honorary professorship of the University of Edinburgh, which seemed to go with the job – and of course a better pension!

In the event, Gerald suggested to the secretariat of the Research Council that they might like to interview Roger Land, a member of Gerald's Department of Physiological Genetics, as a potential candidate for the directorship. Gerald regarded him as having the right qualities. Clearly, Roger made a good impression on his interviewers as he was duly appointed to the post. Gerald was appointed as his deputy, until he retired in 1986. Sadly, Roger did not hold the position for many years, due to his untimely death.

The ARC, in its determination to focus on the more fundamental processes underlying genetic differences in livestock performance, forced a 'marriage' of ABRO with the Institute of Animal Physiology in Cambridge, but it was tolerated rather than liked and was dissolved as unworkable after barely three years.

ABRO was then merged with the Poultry Research Centre in

Roslin, initially under the directorship of Professor Grahame Bulfield. This merger was a logical extension of the collaboration and joint use of resources urged in Gerald's report to the ARC during the 'crisis'. So was born the 'Roslin Institute', which achieved early acclaim and much public recognition for 'Dolly the Sheep' – the first mammal in the world cloned from an adult cell and produced by a team led by Ian Wilmut (later knighted for the achievement).

Little had Gerald anticipated these later successes when Ian had been a member of his department in ABRO. The Roslin Institute, now part of the University of Edinburgh, has become a world leader in the field of animal sciences. The research in its large division of genetics and genomics comes closest to what ABRO had been.

Included in its programme, and greatly enhanced from its early beginnings in ABRO, is important involvement with molecular genetics and gene transfer, largely in relation to disease. It was in support of the latter that the idea of cloning was originally conceived – but that exciting development does not belong to Gerald's story.

Other divisions of the Roslin Institute are concerned with aspects of animal and human health, including studies into the ability of pathogens to transfer from animals to humans, and with elements of behaviour and animal welfare that had been an important component of the studies at the former Poultry Research Centre. With a total staff of close to 400, and large numbers of post-graduate students and visiting scientists, the Roslin Institute must be judged to be outstanding among animal science institutes for its resources, both material and intellectual. So the title to this chapter might appropriately be rephrased as 'From Glory through Turmoil *back to Glory*'.

Glory indeed. On the 25 February 2015, Dolly, the team that had produced her, and the Roslin Institute were accorded another honour by the unveiling of a 'blue plaque' from the Royal Society of Biology. Blue plaques have a long history, but were adopted by the Biology Society only relatively recently to mark outstanding events or persons linked to their location. The one for Dolly was the first to be awarded in Scotland.

Gerald was invited to the unveiling ceremony and was but a bystander, as the cloning had occurred after he had retired. But he was happy to meet and chat with those members of the team who, in their younger days, had been in the department of which he had charge. These were Ian Wilmut, leader of the team, Bill Ritchie, the skilled manipulator of nuclei, and John Bracken, who looked after Dolly's needs. Sadly missing due to illness was Marjorie Ritchie, or Fordyce, as Gerald had known her, who was in charge of surgery. They all looked older and more distinguished but in Gerald's eyes they were still youngsters. He didn't ask his erstwhile colleagues if he looked as ancient as he felt in the surroundings of the palatial new buildings of the Roslin Intitute.

Chapter 28

Retirement, a New Career and a Quick Exit from Yemen

Biggar is a small market town in Lanarkshire, and it was here that Gerald and Margaret settled ahead of their retirement. The town is roughly equidistant between Roslin, where Gerald was still working, and the school in the Clyde Valley where Margaret was head teacher.

The house, a lovely 1870s stone building, had features harking back to the Georgian era rather that to the more fussy Victorian style of its time. It was more gracious a house than either of them had lived in before. It had originally been the schoolhouse for the headmaster of Biggar High School. Like many of the houses of its type and time it was built back to front. The front of the house faced south onto a sizeable garden with views of the hills beyond, while the back of the house, with the 'servant's entrance', the most used entrance to the building, faced towards the street. The garden allowed Margaret to indulge in her love of flowers and Gerald his interest in growing vegetables. Much of their spare time in the first few years was spent turning the garden from something akin to rough grazing into a place of beauty and enjoyment.

In 1986, when Gerald reached the age of sixty – then the compulsory age for retirement – he did not feel ready to put up his feet and contemplate the world. For several of the following years he would drive to Roslin, where Grahame Bulfield, the new director of the institute, allocated him a room of his own and accorded him the status of a 'visiting scientist'. Here, with the help of former colleagues, but most especially of Jeff Lee, he was able to complete much of the analysis and publication of results from his sheep experiments. This work had had to take second place during his years as department head and deputy director.

Following his arrival in Biggar, Gerald joined the Church of Scotland church in the town. He was already an elder of the church from his time in Edinburgh and thus was immediately invited to join the Kirk Session (that is, the group which is appointed to help the minister oversee the life of his church). Gerald was happy in that role, joining in its meetings and serving on the management committee of the Gillespie Centre – the church's local community centre. But the requirement for elders to visit members of the congregation was new to him. In Edinburgh, that duty had been undertaken by a separate pastoral group. These visits were quite a challenge for him as he was a recent 'incomer' to this rural and farming community. Some of the families traced their roots in the district back for generations. Two of those allocated to him were Lord and Lady Clydesmuir, friends of the queen, and the so-called gentry of the area. Lady Clydesmuir, or Joan, as she liked to be called, was widowed within a few years of Gerald's appointment. She was a very sweet, kind lady – a local benefactress, liked and admired by the local population. And her friendship with Gerald and Margaret had an unexpected consequence.

Ever anxious about the life of the community, the indefatigable Gerald initiated the establishment of a charitable organisation to enhance the environment and build on the heritage of the town. The project materialised through the energy and enthusiasm of a committee with Joan Keys as treasurer and Morag Renwick as secretary and with Gerald as the chairman. Many from the community joined as members and volunteers for work on the intended projects. The first of these was in a small area of woodland, long neglected, that was part of Lord Clydesmuir's estate. This is where friendship with Joan Clydesmuir helped. She persuaded her son, the new Lord Clydesmuir, to lease the land to the Heritage Company for a token rent and for fifty years. It allowed the woodland to be redeveloped for use by the community for leisure activities, as a park for walking dogs and as a breeding sanctuary for birds large and small. Volunteers worked tirelessly on this for many months over more than a year. A former local quarry was also developed as a wildlife sanctuary. Sadly, after Gerald left Biggar some of the other initial enthusiasts decided they needed a rest from these time-consuming affairs. By all accounts the Heritage Company is now only ticking over and no new projects have been started. A possible project that had been discussed was the development of a piece of land as a wetland. The land on offer was an area prone to flooding on the edge of the town. That this project could not be pursued was disappointing for Gerald, as there had been much encouragement for it from the Royal Society for the Protection of Birds.

Gerald also served on the community council, for several years as deputy chairman. When eventually the Wieners left Biggar, in 2008, to move north to Inverness, there was a ceremony to thank Gerald for his work, and to present him with a plaque as a memento of his service and a thank you for his work in the town. Theirs had been a happy home where visitors were many, including the grandchildren, eleven in total (seven in the Highlands, and four in London). Gerald and Margaret left behind a beautiful garden, full of roses, shrubs and trees and with a charming summerhouse. On hot summer days, they used to sit there, a water feature outside burbling away as in an oasis.

Gerald had not lived in Biggar for long when, in October 1987, came an invitation from the FAO to go to PDR Yemen as a consultant for the animal component of a project to improve goat and sheep production and fodder crops.

The senior consultant, Phillip, a South Korean, was the crop and fodder expert. He had been in Yemen on many occasions, including a two-year stint. This made him very familiar with the country, which he regarded as his second home. His English was good but far from perfect. These facts assumed some importance later.

As is usual at the start of a consultancy for FAO, there were an enjoyable few days in Rome for background reading and briefing. Gerald found time for return visits to some of the famous sights and to enjoy pleasant evenings in the restaurants favoured by FAO staff. One of the nice customs was that at the end of the meal the proprietor of the restaurant would bring three or four bottles of liqueurs to the table for the guests to help themselves – at no charge to the customer!

On the flight to Aden, Gerald happened to be seated next to Heather, a nurse returning to Yemen from home leave. She had been in Yemen on behalf of the Red Cross for a year or two. On Saturday nights, she told Gerald, British expatriates were invited to the consulate for drinks, snacks and socialising and she suggested that he join that merry throng.

There was little opportunity, outside his work, to mix with local people. Foreigners were largely segregated and he was allocated to a hotel on the outskirts of Aden – a pleasant enough hostelry with outdoor dining in a tree-shaded garden.

The consultancy involved visits to an experimental farm that had been set up as part of a previous project for breeding goats and sheep, as well as trips to other areas with goat and sheep herding. Needless to say, there were also the usual meetings with officials and the local scientists. Gerald's remit was to provide advice on breeding strategies with the aim that improved stock would filter down to the industry at large. It was not to be a project solely confined to the government experimental station.

Saturday nights were the pleasant occasions predicted by Heather, the nurse. During his second week in Yemen, Gerald became quite unwell. He started to have diarrhoea, stomach pain, loss of appetite – and he was losing weight. Heather spotted him at the consulate and thought he looked unwell. She insisted he tell her his symptoms, which he felt shy about – they seemed too personal. 'Listen, I'm a nurse and used to hearing symptoms,' said Heather. Within moments she had diagnosed giardiasis – caught by ingesting food contaminated with the parasite giardia lamblia. She prescribed appropriate medication and by the next Saturday he was already much improved, eating again and regaining weight.

At the start of the fourth week in Yemen restrictions to travel were put on foreigners; they were to stay within an area in and around Aden. Armed soldiers and tanks started to appear at road junctions. That Saturday, the British Consul announced to the assembled company that anyone who did not have essential business was advised to leave Yemen as soon as possible. Gerald's work was almost done, apart from the important task of writing the reports, which usually took a few days. Phillip, Gerald's senior, thought it was time for his colleague to leave. There was a flight leaving for Addis Ababa the next day – the report could be written there. Phillip said he would stay behind and complete the formal meetings with officials. Also, he said, he had many Yemeni friends who would put him up in case of trouble. He said he would send a message to FAO in Rome, though it was likely that it would be intercepted and read by the Yemeni authorities. Accordingly, he decided to say in his message that 'Dr Wiener has left for Addis Ababa for "personal reasons" and will return to Rome for debriefing from there'.

Gerald enjoyed his three days at the Hotel Ethiopia, where he had stayed in earlier years. The hotel had been jazzed up and the female staff smiled and looked as beautiful as ever – though there were no more ladies of the night in the lobby. His time was spent diligently writing a long report dealing with the production of improved goats and sheep as part of the project, but also to provide background to the thinking about genetic improvement (a measure of 'teaching' was part of the consultancy remit).

On returning to Rome, the finance department at FAO refused to pay for the extra flight costs associated with the diversion to Addis Ababa, or to pay the normal salary and living costs for that period. They said that going to Addis for 'personal reasons' gave them no choice – perhaps, they suggested, he had been just visiting a girlfriend. It took some weeks for the project officer, Viv Timon, who knew Gerald well, to sort things out. The money owed was paid in the end. If only Phillip had written to say that his colleague was leaving Yemen to write his report – it would have left the finance department puzzled but not concerned.

One other event from Yemen stayed in Gerald's memory: it was his visit to Shibam, the sixteenth-century city of skyscrapers built entirely of mud bricks and surrounded by a fortified wall. Gerald found walking through the narrow streets awe inspiring. How it had survived as a living town bustling with people and activity was something of a mystery to him. This desert city is now a World Heritage Site and is said to have origins in the second century AD.

Chapter 29

1988 – A Busy Year

After returning from his consultancy in Yemen, Gerald went back to his now unpaid research work at Roslin. He often said to his former colleagues that returning to the institute was for him like coming home. Of course he was also keen to get the results from his former experiments written up and published as soon as possible. Then came a new exciting interruption to that process.

The year of 1988 was to be a busy one for Gerald. First he received an invitation from an Indian breeder, Bon Nimbkar, via the British Council, to go to India for a project vaguely described as to do with sheep and goat breeding. It became for him a remarkable experience for reasons not solely connected with animal breeding. As he recounted later, it is not often in life that one becomes associated with genuine philanthropists. In Gerald's life, this happened twice. First there was his boyhood association with the Spooner cousins in Oxford, who had a great influence in shaping his life. They, as already described, devoted their lives to the service of others and gave their all to that end. The second philanthropist he was to meet and work with, for the project in India, was Bon Nimbkar.

Bon's father, an Indian of high caste, married a white American lady while living in that country. Bon and his wife, Jai, lived in Phaltan in the state of Maharashtra. He was about sixty years old when Gerald first met him and retained his father's high caste – though he tended to joke about that and did not flaunt it in any way. He had held many important posts in his own state in India, but also nationally as, for example, that of Agricultural Director of the Central Bank of India. As part of his and his wife's philanthropic activities they had set up local schools and endowed health care centres whilst themselves leading a simple, almost Spartan, life.

They used their money in the service of others. But Bon's driving ambition, to which he devoted most of his life and great talent, was to help improve the lot of farmers, small farmers in particular, in the state of Maharashtra. To that end he had set up in 1968 the Nimbkar Agricultural Research Institute as an independent public foundation on a non-profit basis. At the time of Gerald's involvement in the late 1980s, there had been sterling work done on agricultural crops and pioneering work on the provision of alternative energy in rural areas. The work on crops had been singularly successful in developing higher-yielding varieties suitable for the prevailing and often challenging farming conditions. A particular example was the development of a variety of sweet sorghum with a sugar content almost equalling that of sugar cane – the then prevailing crop for producing jaggery (an unrefined sugar). But the overwhelming advantage of the sweet sorghum, as for sugar beet in Europe, is that its residues have high value as animal feed, whereas sugar cane has none for that purpose. This alternative to sugar cane was widely adopted.

Thus with more animal feed available from crop by-products and with wastelands turned into better grazing for animals (by introducing different pasture species – another of the institute's projects) the scene was set in Bon's mind to develop improved breeding practices for sheep and goats and perhaps introduce new breeds. Some of the sheep production in Maharashtra, largely with the native Deccani breed, was in the hands of migratory shepherds herding sheep on waste ground and on the edges of roads. Herders, Bon decreed, must be helped to improve their income and the lives of their families. Goats in small numbers were important for village communities by providing milk (as did sheep) and meat. Their productivity too might be improved. And so it was that Bon Nimbkar decided to seek guidance on sheep and goat breeding for a new project at his institute. Gerald's name had come to his attention in this sphere, and so with some financial assistance from the British Council, Gerald was invited in 1988 for an intensive three weeks of consultation. He stayed with the Nimbkars in their home and enjoyed their delightful company and that of their family.

The intensely packed consultation period included visits to various sheep and goat development projects in Maharashtra, to agricultural institutes and the agricultural university as well as to village farms. Some time was also spent alongside migratory shepherds and their flocks of sheep. What an insight it provided Gerald into a way of livestock production so different from that of Scotland or indeed that of most western countries!

In consultation with Bon and others familiar with the agricultural needs of Maharashtra, Gerald was able to provide a detailed set of recommendations for options to improve the genetic potential of sheep, and later of goats. At times, Bon's enthusiasm, exuberance and capacity for generating ideas had to be curbed. Aiming too high could so easily have led to expectations that could not be fulfilled with their limited resources, which would then result in disappointment. Bon had been trained in the USA as a biochemist, and like many Americans was prone to think big. But of course he and his institute had many successes to their credit, albeit in areas more tractable to innovation and change than livestock.

One of the requirements of the proposed programme was to have someone at the institute able to run and develop the programme, with the knowledge of how to analyse and interpret the results of the trials. Chanda, though without previous background in genetics, felt she would like the challenge of that role. Bon thought no-one could do it better than his daughter. Consequently, Chanda enrolled for a postgraduate diploma course in animal genetics at the Institute of Animal Genetics in Edinburgh. She topped the class and was one of the very few students in any year to attain a distinction in this difficult course. In later years she went on to complete a PhD at an Australian University that was by far the most challenging she could have chosen and, in addition, in a very difficult topic. Chanda had not wanted to study for a PhD that might be regarded as an easy ride. She completed her tough assignment with flying colours. The Australian connection led her and her Australian colleagues to track down the mode of inheritance for high prolificacy of a small breed of sheep native to Western Bengal, the Garole. This work led to the discovery and identification of a single gene that held great practical importance. It was found that the gene could be transferred to other breeds, by a process of crossbreeding and backcrossing, to increase their lamb production. It seemed very likely, as a result of later investigation, that this gene had been transferred, in times past, from the Garole to Merino sheep in Australia – a type that came to be known as the Booroola, famed for its high prolificacy. It solved the long-standing mystery how one strain of Merino sheep could be so much more prolific than the others.

Chanda has become well known across the world for her work in this area, but perhaps most importantly she became accepted in India as a leader in her field – no mean feat in the male-dominated environment of that country. Gerald feels pride in having helped not merely to see the sheep and goat experimentation successfully pursued, but in knowing that it has been applied to the benefit of the farmers and herders of sheep and goats in the state of Maharashtra – with ripples spreading further afield. Most of all he admits to a feeling of joy for having been instrumental, though quite by chance, in the emergence of Chanda as a livestock geneticist of note and accomplishment.

Gerald had not long returned from India when, in 1988 also, an invitation arrived from Dr Jasiorowski, a distinguished professor of animal science in Poland and at that time in Rome as the Director of the Animal Production and Health Division of FAO. Jasiorowski wanted Gerald to come to Rome for possibly six weeks as technical editor for a book. The book had been written in order to publish the long-awaited results of a major experiment, mounted by FAO, to find the best strain of Friesian cattle from among the many strains of what was soon to be the most important dairy breed worldwide. Gerald was approached because of his wide experience, having edited an animal science journal and with his own scientific writing and research involvement in this field.

By any standards the experiment itself was massive in scale and ambitious in its objectives. Thirty thousand cows on seventy state farms in Poland had been inseminated with semen from Friesian and Holstein bulls, emanating from ten different countries worldwide. The observations were not restricted to milk yield alone. Also included were milk quality, growth and meat characteristics of the animals, their health and regularity of producing calves and, something important to farmers, the amount of feed the animals used to convert into milk, meat etc. This would provide information on the economics of production of the various crossbred types.

This was a challenge Gerald was happy to accept. He had grown to love Rome on his previous visits in 1972 and 1975. He found accommodation in a comfortable family-run hotel on Aventine Hill. He enjoyed evening meals with some of his pals from former years who had been seduced to join FAO, not particularly by the level of pay but by the feeling that they were doing some good for food production in the developing world. And surely the sunshine and lifestyle of Italy must have played a part. As Gerald had noted before, that 'lifestyle' included convivial evenings in restaurants frequented by FAO staff.

In spite of working exceedingly long hours on the aforementioned book, there was some time in between, and especially on Sundays, to walk around Rome or its outskirts to admire the splendours of both the ancient and the modern city. This is something he was able to share for a short time with Margaret. To Gerald's delight, she had been able to come out to join him in Rome for a week.

Jasiorowski was a larger-than-life figure – tall and broad with a large head and loud voice, and of course a very loud, infectious laugh. When Gerald and Margaret were invited to his apartment in Rome for a small party with his friends they were happy to accept. Jasiorowski's wife seemed uneasy at having these foreign guests. She came into the room occasionally, carrying Polish appetisers on a tray, appearing quietly around the corners of the heavy furniture in this impressive apartment. Her husband seemed to talk endlessly while drinking ice-cold vodka. He told stories to and about Gerald, and then laughed heartily. One that amused him was when Gerald told him proudly that his grandfather had walked hundreds of miles from Poland to Liegnitz in Silesia to find work, just before the First World War, and when he retired he had moved to the big city, Berlin, just before Hitler came to power. 'That was a very stupid thing for him to do, Gerald!' he said, and then he laughed loudly. And when he poured out Black & White whisky, his own favourite, and offered some to his guests and to Gerald, he loudly remarked, 'No black-and-white cows today, Gerald!' This was a reference to the book on black-and-white cattle on which Gerald had been working, but the joke was lost on some of the other guests.

Later on, they had a trip across the city in Jasiorowski's luxurious limousine. Sitting beside this larger-than-life Polish figure as he drove across Rome, with views of the ancient classical buildings floodlit, and past the bright, noisy cafés, made for an unforgettable evening.

In the autumn of 1988 Gerald was back in Rome, this time for a project called, somewhat grandly, 'Strengthening and Developing Animal Husbandry in China's Southwest National Minority Areas'.

It was to be concerned mainly with the yak, the cattle species predominant on the vast Qinghai-Tibetan Plateau of Western China. When asked by FAO to lead this project, the assistance having been requested by the Chinese government, Gerald's first reaction was to say that he knew nothing about the yak and to decline the offer. He was, however, quickly persuaded that his general experience was what was needed and, in particular, that he could be trusted to exercise tact in his dealings in China. Tact was the attribute totally necessary in negotiating with Chinese people. Also, of course, it was likely to be a unique experience for him.

Following the usual briefings and a lot of reading up about yak in the FAO library in Rome, Gerald arrived in Beijing. He was picked up at the airport by a driver from the FAO Beijing offices and, after a brief courtesy meeting with the FAO representative in China, he was taken to a downtown hotel. Still a bit jet-lagged, but more hungry than tired, he made his way to the hotel's large and somewhat gaunt restaurant. The dining room was rather empty and seemed a daunting place to him. Fortunately the menu was printed in both English and Chinese and the waiter smiled. Gerald decided on this occasion not to be adventurous in his choice of food and ordered soup and pork stew with rice. The main course arrived first and with no sign of the soup. Well, he thought, maybe the waiter had misunderstood – after all, the man's English seemed to be restricted to the names of the dishes on the menu. And Gerald had no intention of making a fuss – not with just two words of Chinese in his vocabulary (the Chinese for hello and thank you). The pork stew was fine, the waiter came to take away the empty plates and Gerald was about to ask for the bill when the soup arrived – not a bowl but something akin to a wash-hand basin full, complete with ladle. It was only later that he found out that a Chinese meal often ends, but never starts, with soup. He had a lot to learn – and not just about food.

He was a bit overawed by Beijing in those first three days in

June 1988 – wide streets almost devoid of cars but thronged with bicycles, buildings somewhat drab, and the distances great between them. The walk to Tiananmen Square and the Forbidden City seemed to take forever but, of course, they had to be visited. Taxis were restricted to the front of a few of the largest international hotels, but in any case Gerald always wanted to explore on foot and here in Beijing no one seemed to object or stop him from taking photographs. In fact, every second person seemed to sport a camera. This city was indeed another world.

After briefings in Beijing at the FAO offices and those of the United Nations Development Programme (UNDP), it was off to Chengdu in Sichuan Province in the west of this mighty country, a flight of several hours.

The Southwest University for Nationalities in Chengdu, which trained students from minority nationalities, was to be responsible for the project. It was aided by United Nations funding and overseen by the Food and Agriculture Organisation.

Chengdu, though like Beijing in that it was a city of millions, was yet so different. From the moment he arrived there Gerald felt at home. There were streets crowded with people, shops large and small, avenues with vendors of goods and crafts, trees and flowers at the side of many roads and a university campus that might have been any provincial university in the UK. The first meeting with the project staff, however, was rather formal and stiff, if not awkward. The Chinese team wanted to figure out this man who had been sent ostensibly to assist them in their project but likely to be totally ignorant of what was to come. They hoped he would be on their side.

As it happened, the leader of the project, the principal of the university, was a specialist in the history of minority religions and therefore sadly ignorant of matters to do with animal production, but he was a member of the Party. His deputy, Mr Lin (not his real name), a self-taught veterinarian, effectively became the boss of the operation. He was self-taught because he had grown up during the Cultural Revolution, when other avenues of learning had been closed to him. His main concern, it turned out, was to obtain laboratory equipment for the recently started Animal Science Division of the university, whose principal objective, up until then, had been to train students for administration in the social and political arena. Mr Lin didn't seem much interested in yak. That was to provide the basis for some concern later on.

Then there was Cai Li, a quiet, charming and gentlemanly professor who knew more about the yak, the predominant bovine of the mountainous regions in China, than almost anyone else in the country. Cai Li had a passionate interest in the yak and had devoted his life to their study. Also present at the meeting was a botany professor, and others whom Gerald could not recall, apart from He Wei, a post-graduate student, as bright as she was pretty, who was to act as Gerald's interpreter throughout. She was attached to and working for Mr Lin on his pig-breeding studies. The close working relationship that Gerald developed with her was to provide him with a useful background to some of the issues that arose.

Gerald spent most of his time listening but, as the staff admitted much later, once they had come to accept the stranger in their midst, he turned out not to be as ignorant as they had imagined. He was a person they could work with.

Gerald was to be joined a few days later by an alpine range management consultant, Mel George, from the USA. He was undoubtedly very knowledgeable in his field but had been insufficiently trained in the art of diplomacy, believing that everything can be done better in the USA – not a good attitude when confronted with a people proud of their achievements and rooted in a long and distinguished civilisation. However, his technical expertise and his delight in eating ever hotter food dishes (a challenge to those in Sichuan who stake claim to the spiciest dishes in China) endeared him eventually to most of the student helpers and staff, if not to the bosses.

Gerald tells of a charming interlude away from thoughts of yak, or difficult meetings. On leaving Edinburgh he had been asked by MacNeave, his good friend in Edinburgh, to take a message to a friend of his in China. He requested that Gerald take greetings to the old pastor of a Christian church in Chengdu. Mac, like Gerald, was a member of the Congregational church at Holy Corner in Edinburgh, and the minister who had officiated at Gerald's wedding to Margaret. He was, moreover, a former missionary to China.

It seemed that Mac had known the pastor in Chengdu and they had worked together. Mac's father, also a missionary to China, had actually started the church in Chengdu. In intervening years, during the Cultural Revolution, the church had been burned down, but rebuilt later with financial assistance, it was said, from the Chinese authorities. Gerald telephoned the church's office to pass on the greetings and was immediately asked to come to the church the following Sunday. They would collect him from his hotel.

On arrival, the service had already started. The small church itself was surrounded by a walled courtyard, which was packed with worshippers. A path was made through the crowd, for an embarrassed Gerald, to the very front row of the church. Folk already seated made a space for him. The service was in Chinese, of course, which Gerald could not understand, and the sermon seemed interminable. Yet it was moving to hear familiar hymns sung in so different a language, and some of the formal prayers were recognizable from their rhythm. After the service, over refreshments, the pastor, a small and aged man, was visibly moved and close to tears to receive greetings and news brought to him directly from his old friend Mac. It made Gerald glad that he had been able to fulfil this obligation. After his return home, Mac and his wife were equally delighted to have news from Chengdu.

The time had come to leave the comforts of Chengdu for the mountainous region to the north where the project was to be centred. To the north-west of Chengdu is part of the huge Qinghai-Tibetan

Plateau, mostly above 10,000 feet and rising to around 16,000 feet, interspersed by mountain ranges. It is populated mostly by Chinese minority nationalities (i.e. not Han Chinese) with Tibetans predominating.

Chapter 30

The Road to the Yak Country and Bangladesh

A convoy of cars with several of the professors and the twelve or so post-graduate students in animal science from the university left Chengdu for the high plateau. It was a memorable experience for Gerald, and perhaps also for his colleague, Mel, the American range management specialist – but, if so, he did not show it. He was too blasé about the whole affair.

Gerald wrote a number of letters to Margaret whilst on the trip – some of which did not arrive until he had returned home – and he also kept copious notes, some recorded on his hand-held dictation machine. It is that information which forms the basis for the account of this journey.

In the car with Gerald were Mel, He Wei as interpreter, the professor of botany, and Cai Li, the professor of animal science and renowned expert on the yak. Also present was the Chinese deputy project leader, Mr Lin. For the protection of the health of the party a doctor was sent along to accompany them. It was a large vehicle.

Two other cars followed them. They were for the post-grad students, many of whom were to become unpaid helpers for the project on the farm, Nong Ri, to which they were heading. The journey was to take a somewhat longer route to the farm as parts of the most direct route were blocked by rock falls. Also, the longer route, Gerald was told, would take them to a surprise location, but the staff would not reveal what it was ahead of time.

Their first stop was for lunch in a city called Deyang. As they pulled up outside a restaurant the doctor rushed to the kitchen, ordered kettles of boiling water and proceeded to sterilise everything – plates, bowls, cutlery, chopsticks – and only then allowed them to order food. Apparently, Mr Lin took it on himself to order for everybody, but there was enough mixture for everyone to take a bit of what they liked.

They stayed overnight in a modest hotel in Pingwu – the students were accommodated somewhere cheaper. The highlight of the evening was a party arranged on the spur of the moment by the leader of the town's community after he discovered that there were honoured guests in the town. Gerald described it as great fun with lots of snacks and drink provided. The students seemed tireless, singing and dancing, and the local girls looked very charming, but it was strictly hands off, even for the students.

The following evening they arrived at the promised surprise location, part of the reason for the detour. This was the Jiuzhaigou National Park. Gerald was told that it covered 180,000 acres. They were able to see only a very small part of the park the next day. Basically it was a long, broad valley rising from about 6,000 feet at the base to mountains of 15,000 feet. They saw during their short visit Mirror Lake, shining in different colours because of varying mineral deposits, cascading waterfalls and an abundance of wildflowers, trees and shrubs. The snow-capped mountain peaks in the distance were a magnificent sight.

A visitor centre and hotels, some still under construction, gave them a more luxurious night than the one before. A room was set aside in the hotel to provide the professors, led by Mr Lin, with the opportunity to hold a lengthy, serious meeting about the project in general and the farm and its facilities, or lack of, in particular.

Gerald was glad to have this opportunity to have an in-depth briefing about the project so far.

They left the following morning and stopped for lunch in the small town of Nanping, where the doctor again did his bit of sterilizing. After that they were held up by a rock fall on to their dirt road from the cliffs to one side – a common occurrence, they were told. Many hands, mostly from a nearby village, cleared the path within the hour. Going through the village, their convoy of cars was clearly a curiosity to be stared at.

Looking back on the trip, Gerald's lasting recollection as they passed through the village was seeing newly slaughtered sheep and pigs hanging from rails at the side of the street. These were the butchers' shops. Men stood wielding large knives to cut off pieces of meat for their customers. There were also stalls with vegetables and fruit, and some stalls with lace shawls and clothes also hanging from rails. It was a new world to him.

The next day was spent travelling. The countryside had changed from the lush surroundings of the park to bleak mountainous areas. As the narrow road wound its way upwards the temperature outside dropped noticeably. They had occasional 'comfort stops' but comfort, Gerald recorded, was for men to go into a field on one side of the road and women on the other.

Soon, as they came around a bend in the road beside a huge boulder the size of a small house, Gerald saw his first group of black, hairy beasts. He thought they were yak but was corrected by his Chinese friends that these were crosses of yak with the local yellow cattle, and called Pian Niu. With his ignorance thus exposed he kept his mouth firmly shut until much later, when he had learned to detect the small differences in appearance between the hybrids and the pure yak. Mel meantime quizzed the botany professor about plants seen during 'comfort' stops and Latin names were bandied about that He Wei struggled to translate.

In the late afternoon, after a tiring drive, they reached the main town of Hongyuan County, where the farm was situated. The hotel was a welcome haven for them, and adequate, rather than luxurious.

After a comfortable night and good meals they set off westwards towards Nong Ri farm, some thirty miles away. Mr Lin, with a couple of the students, had streaked ahead of them. The countryside now changed to something very different from what they had seen on the journey thus far. A vast plateau stretched out before them with a wide river and smaller streams winding through it. The land was covered, it seemed, with hundreds, perhaps thousands of black, hairy animals – this time the real, pure yak. Gerald described them as looking different from buffalo but also different from cattle – perhaps something in between. Alongside the yak were also white sheep. It was a lovely sight, but Gerald was quickly told that this area had one of the densest populations of yak in the country and that it was not typical.

When they arrived soon thereafter at Nong Ri, along a relatively good if narrow tarmac road, their first sight was of Mr Lin up a telegraph pole, waving to them. It was Mr Lin's idea of a welcome.

They stopped outside some grey concrete buildings, which were the headquarters of this local-government-sponsored farm. The manager explained that from there they administered an extensive area of plateau and hill land apportioned to a number of families of yak herders, each with their own herds. This was also the central location for the intended investigations involved in the yak project.

There was a small hostel for their accommodation, a dining room and, specially built for the two western guests, a small cabin with a shower and toilet. Sadly, it did not always function. Gerald said later that it was nonetheless an improvement on the communal wooden toilet building, which had about twenty holes above a cavernous pit. That day ended with a meal, mostly of vegetable dishes, boiled rice and a small amount of diced meat – not as elaborate as meals of the past three days. Gerald and Mel shared a room.

The following three days were spent on and around the farm. The animals for the project were to come from six of the herder families, with a total of 900 yak at their disposal. A representative sample from each herd was to be made available for trials and experiments and the herders were to be paid for the use of the animals. The site allocated for the project's work consisted of about 500 acres of fairly flat valley grassland, which was regarded as the winter pastures. The total area of Nong Ri farm was vast, about 40,000 acres, with more than 9,000 yak in total. Some of that land was marsh and much of it hill country utilised for the animals in the summer months. With eyes half shut, Gerald claimed he could imagine himself back in parts of the north of Scotland, until he remembered that he was at 11,000 feet and the hills rose about 3,000 feet above that.

There was much discussion of the facilities that would be needed. A crush with weighbridge was required to hold animals individually so that observations could be taken. A small laboratory with refrigeration was also required for examining milk and storing blood samples. Electricity on the farm was generated from a mini hydroelectric scheme cleverly constructed in a nearby small stream. Mel also wanted a little of the adjacent land fenced off into fields for potential grazing and feeding trials.

Some distance from the 'headquarters' section of the farm were the dwellings of the herdsmen and their families. These were large, black tents, most of them made traditionally, from yak hide, but a few from conventional tent material. As this was late summer, most of the herders were still with their yak on the summer pastures higher up in the hills. These were grazed in rotation – it was called a transhumance system of herding (not the full nomadic style).

The day prior to departure, the party was invited to visit the habitation of one of the herders and his family. Gerald described that visit in some detail. They drove perhaps half a mile to the tent where one of the herders lived with his family. He owned around

100 of the yak that were to be made available to the project. They were told that he had come down from the hills especially for this occasion. The large black tent, with smoke coming out from a hole at its centre, was guarded by two large, ferocious-looking dogs. Not, as it turned out, to keep people at bay but to deter wolves.

Outside the tent was a large stack of bricks. These turned out to be slabs of dried yak dung, which was the main fuel available to the families given that there were no trees at these elevations. Inside, they were welcomed by 'the boss' and perhaps ten of his family members. Women and children were seated on couches around a central fire. Suspended above the flames was a large black pot with a strong brew of milk tea bubbling inside. This was, they were told, a favourite brew in the yak territories; the milk was yak milk, which is very creamy and slightly sweet tasting. Also on offer were pieces of barbecued yak meat and Gerald, as the chief honoured guest, was offered a roasted yak jawbone, complete with teeth. Gerald knew that it was considered *de rigeur* to show appreciation and accept what is offered, but he drew the line at a jaw with teeth. He thinks that the family had not really expected him to sink his teeth into the yak jaw, as they smiled gently at him. Mel refused to take anything, even the tea, and Gerald suspected correctly that this was not good manners anywhere, but close to an insult in this situation.

The next day the party set off down the shorter route back to Chengdu, crossing a pass at 14,000 feet. The party got out of the car to see the glorious vista of mountain grazing land with yak herds in the distance, but the air was too thin to linger long.

Back in Chengdu and at the university for final discussions, it became obvious that there would be a conflict between what Gerald and Mel regarded as the needs of the project and the desire, voiced mostly by Mr Lin, to equip laboratories for the students and staff of the university. The impasse was carried over to a final meeting in Beijing with senior project staff, the consultants, the head of the FAO mission in China, and a senior official of the

Chinese Government. Gerald, practising diplomacy to the best of his ability, proposed that he would request that next year his mission would include an expert in laboratory equipment to assess the needs and relevance of the requests. Failing an agreement on equipment, he said, he would try to persuade the FAO to grant additional fellowships for post-grad students to study abroad, in place of extra equipment. This, Gerald suggested, would be a valuable investment for the future. The suggestions got a nod of agreement from the FAO man from Beijing and the meeting ended amicably, though Gerald doubted if Mr Lin was happy.

So ended Gerald's first trip to China. Mel left directly for the USA and Gerald returned home via three days of debriefing at the FAO in Rome. But nothing was as good as the return to Biggar and Scotland.

Before returning to China for the next yak consultancy visit, Gerald was asked by the FAO in early 1989 to go to Bangladesh for a project not, for once, related to any specific livestock species but concerned with broader issues.

The project to which he was assigned in that impoverished country was to provide assistance to its still young Livestock Research Institute in developing its programme of research and to enlarge its limited facilities. This did not require Gerald's experience as an animal geneticist, but it did draw on his understanding of the broader issues of running a research department and institute, combined with knowledge of animal production and the requirements for experimentation.

As he recalled, the task was far from easy. To quote from Gerald's report:

> *The work undertaken (at the Institute) seems to reflect more the specific interest of the research division chiefs and possibly their staff, as well as the directors.*

> *There was no coherent strategy towards solving relevant problems. These should have been restricted to a small number of high priority areas where the limited resources of the Institute might reasonably expect to make an impact in a reasonable period of time.*

Moreover, Gerald found, on first examining the work being done there, that such projects as had been undertaken were often too small in scale to provide credible results, were inadequately designed, and that the titles assigned to the projects were more all-embracing and made to appear more significant than the content of the work justified. This is where tact was needed in plenty. His major concerns had to be addressed with respect to the problems that faced an under-funded and under-resourced institute with staff that were still relatively new and inexperienced. But in that spirit, new directions of experimentation were agreed with the local staff and their directors.

Part of the time in Bangladesh allowed visits to villages where farming, in its broadest sense, was the most important enterprise. Apart from poverty and poor resources, the clear impression was that everyone worked hard and made the best of very little. In particular he noted the critical role played by women, especially in tending livestock and utilising their products. As far as livestock were concerned the importance of poultry in the local community stood out.

Being a confirmed sightseer, and interested in the general life around him, Gerald was always keen to visit the historic parts of any city. He wanted to see the markets, the shops, the schools and the entertainment. Bangladesh was the only one of the countries he visited where he was unable to pursue this interest. A tall perimeter fence surrounded his hotel in Dhaka, and the entrances were guarded. Outside that fence was the constant presence of great numbers of destitute children begging. It was a heart-breaking sight and one that contrasted with the relative luxury of the Hilton Hotel. Never was he able or advised to venture outside on foot. Each morning and evening a car would take Gerald to the institute and then drop him back off, and whenever the car slowed down or stopped, it would be surrounded by the children. These sights made him wonder how soon, if ever, the international aid for research and development to agriculture, fisheries and forestry would make an impact on the lives of the people – and whether his own tiny role would be of any use in the wider context.

CHAPTER 31

Return Visits to China

There were to be several returns to China for Gerald over subsequent years. The first of these was in August 1989 to re-engage with the yak project. After a brief time at the FAO in Rome, he arrived in Chengdu at the Southwest University for Nationalities, which had the responsibility for running the project at Nong Ri farm on the high plateau.

The reception he received differed markedly from that of the previous year, when he had been met with courtesy but also great reserve and some suspicion. That seemed to have been dispelled and now he was welcomed as a colleague. They say that it takes Chinese people a while to get the measure of strangers before they will accept or trust them. Perhaps that would be a good idea everywhere. On this occasion Gerald was accompanied by two new consultants. Mel had been replaced by John Frame, a grassland and clover expert from Scotland. This had come about because the specialist officer at the FAO headquarters in Rome had not agreed with Mel that what had to be addressed was a range management problem – his personal speciality. The experts in Rome thought that the path to better nutrition for yak, and sheep, lay with introducing new grasses and perhaps clovers. This proposal failed to take account of the fact that not twenty miles from the Nong Ri yak station was the Sichuan Provincial Grassland Research Institute. It had been experimenting for years with different species of grasses and clovers to discover which of them could be established in these harsh regions. John himself doubted whether reseeding areas of grazing would be feasible, but he was happy to join the team. He was a very pleasant companion to have around.

The third member of the team, Alex Field, had been appointed as an animal health consultant with expertise in laboratory equipment. This had followed Gerald's request for such a person following the insistent demands from Mr Lin for the equipment. Alex was a former long-time colleague and collaborator of Gerald's in his research on the influence of heredity on copper metabolism in sheep. Thus a threesome from Scotland appeared in Chengdu in August 1989.

The visit to the project site at Nong Ri farm was eagerly awaited. The experimental area had been fenced off and different paddocks created. Facilities for restraining, weighing and examining animals had been created, but these were short of what they should have been. As an example, the scales for weighing animals were of a very crude type. Without greater precision it would be difficult to detect, with sufficient accuracy, the effects on the yak of different feeding regimes. At that stage in the project the university staff were not yet accustomed to the rigour necessary in conducting trials, or in the requirements for statistical verification of results. The participating herders had also put a slight spoke into the wheels of the project by not signing the agreements for the use of their animals. They were holding out for better payment and had demanded television masts to be erected so that their community could receive television. No one had anticipated that the simple yak herders would be keen businessmen. In consequence, few actual observations on yak had started, but these were now promised.

The emphasis was to be on collecting as much factual information as possible on all aspects of performance of the yak, as a prelude to any future improvement programme. There were also to be trials involving different methods of winter-feeding of yak cows to see how this would affect their subsequent performance and the survival and growth of their calves. In the main, the yak, even pregnant females, came close to starvation from mid-winter through to around May, when the first growth of herbage starts to appear.

Trials at conserving fodder for the winter were also planned. Giving small amounts of extra feed in winter was traditionally restricted to sick animals. Experimental introduction of new grass species could be done collaboratively with the Grassland Institute down the road. All of this was a big learning curve for the staff of a university good at training administrators but with no real experience of field experimentation with animals. The vexed question of laboratory equipment arose again, but this time in the presence on the team of an 'expert', who, with difficulty, received grudging respect from his Chinese hosts. Perhaps the most pleasant duty for Gerald and his colleagues was to assist the national project director (i.e. the head of the university) in selecting post-graduate students for FAO fellowships for study abroad.

There was one further interesting episode from this consultancy tour. The FAO had introduced the idea that, in addition to international consultants, there should be national consultants attached to a project. These Chinese consultants might keep an eye on the project and provide additional advice and support throughout the year. For this purpose two of the top men in their field in China had been identified by the FAO. One, Professor An Min, an internationally known animal scientist and former head of Beijing's Agricultural University, was retired. The other was Professor Ren Jizhou, who was the director of the top Grassland Institute in China. He also had an international reputation.

However, the people in charge of things in Chengdu were fiercely opposed to what they saw as interference with their supervision of the experiments. Gerald was told that the men chosen, An Min in particular, were 'too old' for the task and in any case 'not available'.

Gerald was disappointed at this decision. By a total coincidence he received word one afternoon, whilst in discussions with the project staff, that the said An Min happened to be on a lecturing visit in a town some twenty miles from Chengdu.

Gerald immediately made a direct request to the local authorities to be taken to meet An Min. Such a request could not be refused. An Min turned out to be charming and perceptive, not too old to withstand the rigours of high altitudes and more than willing to accept the challenge alongside Ren Jizhou. It was slightly harder to convey this 'coup' to the national project director and his deputy in a way that would not make them lose face. They had, it turned out, never been in touch with An Min and his refusal to accept the consultancy had been an invention. Gerald had learned in his previous work how to turn what might be seen as a rebuke into something resembling a compliment, though it was somewhat difficult to achieve on this occasion.

'Team Scotland' – that is, Gerald and his two colleagues – returned to Chengdu and the yak project in August 1990. They travelled together for a briefing at FAO headquarters in Rome. Happily, there was again enough time to take in a little of the ancient grandeur of that city and some of the wonderful Italian food. Alex and John were especially delighted, as Gerald had been earlier, by the bottles of liqueurs put on the table at the end of the meal with an invitation to help yourself as it was 'on the house'. The three arrived together in Beijing and then travelled to Chengdu in happy mood. Little did they know how this trip would end.

On his return home Gerald was asked by some of his former colleagues at the Roslin Institute to give a talk on the yak project. Since he always prepared his lectures and talks in writing, and kept a copy, an extract about this third visit can be given in his own words – leaving out all the detail of background and technical aspects of the project.

At last there were some actual observations on the yak at Nong Ri to look at. A better weighbridge had been installed and some results on weight changes over the year and estimates of milk production and composition were available. This counted as progress. Also there had been two winter-feeding trials. One was to use hay over the worst part of the winter and early spring. The yak cows so fed were in much better shape by the time their calves were born than those without this supplement. About half the cows produced calves and they too benefited from their mothers' winter-feeding. This may not seem surprising here in Scotland but for the yak herders at Nong Ri it was a new experience.

However, the hay had been very difficult to procure, as the herders had had to painstakingly cut unevenly growing herbage on very uneven ground. They did this with sickles, as something larger, like scythes (that I had wanted them to obtain and use), would not have worked on this terrain. Thus, making hay in sufficient quantity for the trial turned out to be very expensive in labour. The good effect of feeding hay was therefore more of a demonstration than a practical solution to the yak's severe loss of weight and condition over winter. But with research in progress at various institutions, the hope was that, in time, some new fodder plant might be found that would grow sufficiently during the short summer season.

A second winter-feeding trial was to have been a comparison of animals given access to standing wilted herbage in a large paddock, fenced off during the growing season, with animals allowed to roam on previously grazed areas. It was then that the real problem of conducting trials in these conditions came to light. Wooden posts had been used to support the wire fencing. But, you've guessed it already, wood is a precious and rare commodity in these territories above the tree line. So many of the wooden posts had been purloined as firewood and yak had wandered in and out of the paddock at will. So no credible results emerged.

Time came for the final meetings in Chengdu, attended, as before, by a senior official from the Ministry of Agriculture in Beijing, and also the head of the FAO mission in China. The project team from the university as well as the three consultants were seated around a massive table – with me flanked by John and Alex – and the two 'high priests' from Beijing at the top of the table.

The etiquette of these meetings is that, in presenting the progress of a consultancy visit, the only ones to speak are the project director – as I told you earlier, the appointed director was a professor of the history of minority religions – and the senior consultant from the FAO side, that being myself, Gerald Wiener. Things went reasonably well in spite of some claims from the project director that I had to dispute and correct.

Everything was quite amicable until the question of equipment came up and the deputy project director, the Mr Lin who I referred to earlier, spoke up saying that they still needed a blood-chemistry analyser. As I had requested an expert in equipment to join our team, because of the earlier pressure for laboratory equipment, I felt it only sensible that Alex, appointed to that task, should respond. What I did not anticipate is how Alex would proceed. Many of you know Alex and his reputation for what is politely called 'plain speaking' – as often practised by him at meetings and seminars here in Scotland. His behaviour in China would be different, or so I thought, but unfortunately, Alex did not think that way. 'You don't need this equipment,' said Alex, 'and even if you had it you would not be able to look after it'. 'That would not happen in China' – a somewhat tetchy retort from the man from the ministry. 'Oh, but it does,' came back Alex in a flash. 'I visited your main hospital in Chengdu yesterday and the very same analyser that you want has stood there broken down for the past two years.'

After a moment of terrifying silence the man from the ministry rose and walked out from the meeting. The top officer of FAO in China was infuriated and stormed round the table and told me that I had just ruined years of good relationship between China and the Food and Agriculture Organisation of the United Nations. He said that I had imperilled all their projects. I am not to be allowed to return next year to produce the terminal report on the project – usually reserved for the senior consultant. So ends my time with the yak.

Well, perhaps this is not quite true. FAO in Rome had asked if I would collaborate with Professor Cai Li – the yak guru at the university – in publishing an English-language book on the yak. Cai Li was keen on the idea and he and I parted as friends, in spite of what had transpired at the meeting, with promises to communicate with each other. In fact, several of the staff, apart from Mr Lin, bade me and John a more than courteous farewell, so maybe there will be another story in years to come.

So ended Gerald's talk to his colleagues in Edinburgh about his Chinese experiences among the yak on the high plateau.

Gerald was in fact asked back by the Chinese for lecturing and conferences and was made an honorary professor of a leading agricultural university in Gansu province and of the Southwest University for Nationalities, the home of the yak project, which had grown to be ranked number six among Chinese universities (how the ranking was arrived at Gerald never found out). So clearly the Chinese themselves had seen through the charade of that infamous meeting and attached no blame to him. It transpired later that Mr Lin was moved sideways into an administrative job away from any involvement with yak – or pig reproduction, the field that he had claimed for himself but which had been found wanting.

Chapter 32

A North Korean Interlude

Not many people were given the opportunity to visit Democratic People's Republic of Korea (North Korea) in the early 1990s, a country largely isolated from the west. At the time of writing the country is still suspicious of the outside world – apart from its principal supporter, China. For decades it has been ruled by a hegemony of dictators, worshipped by the masses as 'great leaders', and supported by a large military force and a communist-style government.

So it was with a mixture of interest, excitement and apprehension that Gerald set out for Pyongyang in November 1991. The usual briefing at the FAO in Rome had stressed more than normally to 'keep off politics, avoid any criticism – except if related directly to your project – and try to admire what you see'.

True to the last of these pieces of advice, Gerald went out on a walkabout of Pyongyang, once he had slept off some of the tiredness of the early morning arrival of his flight. There were no work-related appointments until the following day, so he set off happily.

Pyongyang was an impressively modern city with high-rise blocks of flats, somewhat dull to the eye, but functional. There were wide streets, a river to brighten the city, many monuments and public buildings to impress – or overawe. Gerald set off, camera at the ready, to photograph many of these sights. At no time was he impeded, but neither was he approached by any friendly members of the public, as might have been the case in China, or in any country where foreigners are a curiosity but the locals don't feel restrained.

Next morning came the first meeting with a senior official of the ministry responsible for the goat project Gerald was about to be involved with. First question: 'Well, Dr Wiener, how did you enjoy your walk around our city?' It was obvious that he had been followed discreetly throughout the previous afternoon. This was a courteous reminder that North Korea expects foreigners to behave and not to pry.

The project itself was concerned with goat production following the establishment, in an earlier phase of United Nations support, of a pilot goat farm with a large area, in excess of 2,000 hectares, at its disposal and projected to reach double that size in due course. So this was no mean enterprise. The purpose was to breed cashmere goats imported from Inner Mongolia and China in order to develop cashmere fibre production in the country. It was as yet unclear whether this was to be restricted to fibre for export or to start a local industry.

Gerald's job was to assess to what extent the initial aims of the importation had been met and to assist in suggesting a future strategy for the cashmere industry. The initial aim of breeding to increase the number from the imported 350 females had in fact been very successful. Over a period of just three years the numbers had increased to around 2,000. That was clearly a credit to the management on the farm. But the sad fact was that Korean officialdom regarded this as failure, because a document at the start had suggested a total of 10,000 after three years. Gerald pointed out that this was a biologically impossible target, by normal reproduction. He never discovered whether the bureaucrats accepted his assurance or continued to believe the proposed figure – however ludicrous.

A second component of goat production was the decision by North Korea to import Saanen goats, a breed renowned for milk production and of a good size for meat. This was to be introduced over time to improve the strain of local goats in villages and provide extra milk, meat and income to the smallholders. Gerald's role, in concert with the local experts, was to suggest various breeding and management strategies for these developments. This was something with which he was familiar. Gerald always stressed that goats were a species highly successful in exploiting sparse conditions in which cattle would struggle. Goats were also much less expensive to buy and to maintain and were thus very useful for rural communities.

There were several visits made by Gerald to the goat farm for discussions with farm staff and with the scientists involved, and for the purpose of becoming familiar with this facility. For the final visit to the farm, before his departure, the staff decided to celebrate and to give Gerald a suitable send-off. In consequence a goat suffered an untimely end, was slaughtered and barbecued on an open fire. Unfortunately, the time between slaughter and plate was too short. As the honoured guest, Gerald was plied by the manager of the station with the largest and best pieces of meat, still only half-cooked. Some pieces had to find their way, surreptitiously, into his pocket. But it was a jolly occasion and perhaps the only time during his relatively short stay in North Korea that Gerald saw genuine smiles and heard real laughter. So perhaps away from the city, the military parades, the unseen police and the staged displays of enthusiasm, the people were like any other – welcoming to guests, friendly and willing to have a good time.

One free afternoon was spent, at the request of his hosts, in the sombre military museum which extolled the country's might. In the evenings – after time on the farm, at research stations and in meetings with officials and with both technical and scientific staff – there was little life outside the hotel, which was reserved for foreigners. Gerald's time was enlivened by Brian Barclay, an economist of Canadian extraction. They had already met on the plane bringing them to Pyongyang and his sense of humour and company were a fine stimulus. His was perhaps a politically more sensitive assignment, and he was taken aback at his first meeting with ministry officials when he was asked why he had not taken part in Gerald's walkabout. Clearly that walk had caused a bit of a stir.

Brian and Gerald continue to correspond to the present day. Brian's witty Christmas letters are suitable for any humour magazine. Whether Brian's expertise or Gerald's had any lasting impact in North Korea is unfortunately something the consultants did not find out.

CHAPTER 33

Return to Berlin

Gerald's lifelong friend Hardy Seidel phoned him from his home in London one day in 1992. He and his wife Ruth had recently spent a week in Berlin, where he and Gerald had been at school together as young boys before the war. It seemed that the Burgermeister of Berlin had issued an invitation, apparently open to all former refugees who had lived in Berlin, to visit the city at its expense. Hardy had taken up the offer and now suggested that Margaret and Gerald might also like to avail themselves of such a free holiday in Berlin. They would only have to meet their airfare. Living in Scotland, and not associated with any Jewish or refugee organisation, Gerald had been totally ignorant of the scheme and the generous wish of the Berlin authorities to rehabilitate themselves in the eyes of the former Berliners.

Margaret wasn't keen. She still harboured suspicion of Germany and Germans – even though she had married Gerald who, though feeling fiercely British, had been born in that alien land. Gerald had doubts too, particularly about how he might react to a place and an environment that he had left behind long ago, both in body and spirit. And he could not forget that some of his family had perished in concentration camps. However, in the event, curiosity prevailed plus, no doubt, the attraction of a free holiday. They flew to Berlin at the end of March 1993.

A quite luxurious hotel had been booked for the couple for a week in a part of Berlin that Gerald must have known well in his childhood but had largely forgotten – until, that is, memories started to flood back during the stay. The day after their arrival, the couple had an appointment at the Burgermeister's office, where they were given an itinerary for the week. Most of the time they would be free to go wherever they wished, with plenty of suggestions for things to see and do. One visit, however, was to be 'compulsory' – to the memorial that had been created for those many Jews and other citizens hounded by the Hitler regime. The building had been a place of torture and execution and had been restored to represent that place of horror. On this visit Gerald and Margaret were accompanied by a guide. The experience of standing looking at the place was both moving and very chilling.

Other parts of their stay were more cheerful and enjoyable. The couple were presented by the Burgermeister's office with expensive tickets to an opera, *Aida*, and to a ballet, *The Three Musketeers,* and given money for their expenses outside the hotel. This was a generous gesture by the Berlin local government.

Parts of the Berlin Wall separating the west and east sides of the city were still in place, covered in graffiti. It was easy to imagine the day of liberation for those in the east when the wall started to come down and people flooded through an open gate, many to be reunited with family members from the other side. Happily, Gorbachev, the leader of the Soviet Union, as it was then, chose not to interfere with the uprising in East Berlin. Reunification of the two parts of Germany followed just a year later, unexpectedly early. By the time Gerald and Margaret arrived it was still easy to appreciate the relative poverty and restriction of those living in the former communist east of the city compared to the growing wealth, freedom and cosmopolitan life of those in the west. One could still see some run-down and unkempt streets and buildings in the eastern part, on the other side of the Brandenburg Gate; in contrast with the bustling crowds and cafés, department stores and extensive rebuilding in the western part. One exception to the drab surroundings of East Berlin at that time was Unter den Linden, the wide avenue stretching eastwards from the famous gate, named after the lime trees gracing its length. It was still a street of grandeur with its opera and comic opera houses, museums and other majestic buildings. It would take many years before the economic gap between east and west was repaired.

It did not take long, however, for Gerald to revive memories of his early life in Berlin: life on Bamberger Strasse with his mother and grandfather; the childhood visits to the KaDeWe, the huge and luxurious department store in the west end of the town, the largest department store in Europe, it was claimed. Its food department, with carp and pike swimming in large glass tanks, had been his joy as a child. The shop had been restored and enlarged after wartime damage and ownership had been returned to the descendants of the original owners, who had been ousted on Hitler's orders because they were Jews.

The apartment block where Gerald had lived with his mother and grandfather was gone – whether as a result of bombing or because of the road widening, he could not tell. The square with its fountains where Gerald and Hardy had played with their toy cars was still in place, perhaps restored. Gerald never found his way back to his old school (as Hardy had done) but they did visit the street where his grandfather had sought retirement in an old people's home after he had been left on his own. But it had not prevented this non-political and harmless old man from being taken from the home to a concentration camp, where he died.

Margaret and Gerald spent some time in the Tiergarten, a lovely park and the scene of many Sunday outings in Gerald's childhood. But alas, there was no time to go to the zoo at one end of the park – a place of great wonder for a child. He could not recall if it had been the elephants or the giraffes or the monkeys that were the biggest attraction for him and for his cousin Marion.

Another childhood Sunday outing had been to one of the many cafés on the Kurfürstendamm. That ritual was repeated by Gerald and his wife, but they did not hear little orchestras playing to the guests, as Gerald remembered. Perhaps they were no longer in fashion.

Apart from the uplifting outings to the opera and the ballet in splendid theatres, the pair visited some of the museums of East Berlin. The Burgermeister's office had also suggested a visit to the new Jewish synagogue, but this idea was not followed up as Gerald had left the Jewish faith behind and curiosity did not seem an appropriate reason for attending a service. Gerald and Margaret did, however, spend some hours at the Kaiser Wilhelm Memorial Church near Berlin's fashionable Kurfürstendamm. The church had been all but obliterated during the bombing of Berlin in World War II. Some ruins and much of the tower remained and were incorporated in a new, modern building as a reminder of the horrors of war. But this was no self-pitying memorial. The large iron cross was made from nails from the bombed and all-but-destroyed Coventry Cathedral in England. This bore witness to an awareness of suffering from war by all sides.

A large exhibition in one of the halls that were part of the church complex showed, side by side, the wreckage wrought by the bombing of Berlin and of Coventry. This visit was a most moving experience for the pair and showed the serious attempt being made by Berlin for reconciliation.

A day was spent on a trip to Potsdam on the outskirts of Berlin – the venue of the end-of-war agreements. The object of this visit was to see the Sanssouci Palace, built for Frederick the Great. This building was magnificent in its interior and incredibly splendid outside, with its terraces and almost magical fountains in a gracious park.

The time for departure from Berlin came all too soon. But first, a visit to the offices of the Burgermeister was called for, to say thank you and talk a little about the week that had passed. In a letter of thanks that Gerald wrote to the Burgermeister for their generous reception, he also gave some of their reactions to the trip. He mentioned the reluctance they had felt about coming and their earlier doubts about the likely success of the experience, in view of some bitter memories Gerald had been left with. Soon thereafter Gerald and Margaret received a reply signed by the Burgermeister. It is of interest to quote from that letter:

> I can well understand the mixed feelings you had about making this journey and I am therefore all the happier that you came, and above all that in the end you were not disappointed. We hope that we were able to contribute personally a little to that, so that you learnt, as far as that was possible, something of our circumstances. If so, our intentions were fulfilled. I also want to thank you for your remarks and good wishes, showing that our discussions were not just superficial. I hope that we shall hear again from each other in the future, and from our end, we shall send you a new publication *Aktuell*.

Aktuell, with information from and about Berlin, arrives at their home in Scotland every three months. Translating it from the German is a bit of a struggle for Gerald, but the pictures and many of the historical articles are splendid. Often this is accompanied by a message addressed 'Lieber Berliner' (Dear Berliner).

Chapter 34

Back to India

The Nimbkars invited Gerald back to India in 2000, not specifically to provide advice on the projects he had helped to start but to allow him to see first-hand the progress that had been made. It was also to give him a chance to further explore their country, of which they were very proud. Of course they included Gerald's wife, Margaret, in the invitation and as this was to be a private visit the couple would be with them as their guests.

Shock is the only word that can describe what Gerald and Margaret felt as they emerged from the airport at Mumbai. They had left behind in Scotland the cold air of Edinburgh in January of 2000 to be delivered, tired and excited, into the heat and darkness of midnight in this great Indian city. There seemed to be hundreds of people, their eyes glinting in the street lights and all their activities illuminated by the passing cars. Taxis were lined up, it seemed in great numbers, and the drivers crowded together around the European travellers, calling out for their business.

It seemed that in no time they had been loaded into a cab and were driving through the ghostly-lit streets of downtown Mumbai. The sights left them silent with astonishment. Whole families were squatting gracefully on the pavements, gathered in small clutches around little cooking stoves. Some stared up in surprise at the passing taxis, their gazes lingering.

To the new arrivals, the air seemed heavy with cooking aromas of spices and other eastern fragrances. There was an eeriness about those streets, especially to Margaret, who had never been to the East.

Later, after a train journey, they found themselves in Pune (also known as Poona) where they were met by Chanda Nimbkar. Having been friends when Chanda was studying in Edinburgh, she and Margaret were delighted to meet again. They bought some fruit and then drove to Phaltan, to the large farmhouse where Gavan, Chanda's Australian partner, was waiting for them. Tea in the front garden under the trees was the start of their Indian adventure. Margaret had brought a pot of homemade Scottish blackcurrant jam in her luggage and they ate that with their Indian bread.

The next day, they were shown the whole farm, accompanied by Turpie, the family's pet dog. As they walked around the grounds, a dozen small sheep passed in front of them, followed by an old shepherd. The sheep were identified as Garole sheep, the breed that had become important because of its high prolificacy. Turpie seemed to enjoy being up close to the animals, poking his nose in their ears. The farmhouse was a handsome, tall building. The ground floors were of stone, and the bedroom where Gerald and Margaret slept had a primitive shower room. They passed the two ladies who cooked and did the housework, who today were hunkered down at the side of the fast-running burn, washing the clothes. Nearby was a device by which methane gas was obtained from sewage. This gas was used for cooking in the house. There were four water buffalo tied up in an open-sided barn. They provided the milk for the household, some of which was made into a delicious, rich yogurt.

Further down the path they reached the fields of sorghum, grown for making sugar. At the edges of the field, colourfully dressed women stood, it seemed all day. Their job was to clap their hands if birds appeared above them. They would give loud calls and clap their hands loudly to chase the birds away. In the heat of the day this was surely a boring job!

Chanda pointed out the jaggery building where they made the sugar. They walked on further and saw more little sheep surrounding a half-asleep shepherd. Ten feet above him, between the branches of a tree, hung a great, egg-shaped swarm of dark bees. These were killer bees, and they were told to be very quiet so as not to disturb them – they are not called killer bees for nothing.

Walking back to the farmhouse, they came upon Gavan dressed in a brown lab coat and attending to some poultry housed in an enclosure among the trees. Gavan was proud of these birds and the house he had constructed for them – so well insulated against the heat that the birds were comfortable inside even at the height of summer. But what intrigued Gerald most was what Gavan, with genetic advice from Chanda, was trying to do. His aim was to recreate, from the local poultry of the villagers, the colours of the red and the grey junglefowl that could sometimes still be found wild. The local chickens, he had observed, had hints of junglefowl colouring in their plumage. After only a few generations of selection and backcrossing he had produced birds indistinguishable from the red junglefowl, but the grey ones were more difficult to recreate. It brought back memories to Gerald from fifty years earlier, when he had worked, albeit as a labourer, with the poultry at the research farm in Cambridge. There, his geneticist boss, Michael Pease, had used similar breeding techniques to produce a colour difference in the birds – though for the entirely different purpose of distinguishing male chicks from female soon after hatching.

As they were leaving the poultry enclosure, Gavan told them of the predators the poultry had to contend with: jungle cats, jackals, eagles, mongooses and even nine-feet-long pythons. But the poultry were his hobby, and the eggs were used in the kitchen. Gerald and Margaret did not care to hear more about the pythons.

Meanwhile, Gerald was keen to revisit the agricultural research centre established by Chanda's father, Bon Nimbkar. He also wanted

Margaret to meet this man and his wife, who were so revered in that part of India for their philanthropic community work. The work on agricultural crops and renewable energy had continued apace. But since Gerald's advice to them on his first visit twelve years earlier, and with Bon's drive and enthusiasm and Chanda now in charge of the breeding work, they had made good progress in their aim to improve productivity from sheep and goats.

There had been some importations of milk sheep from the Middle East and Boer goats from South Africa, and there had been the significant discovery of the potential of the Indian Garole breed to boost lamb production in other breeds. But the most important progress was the start that had been made to transfer improved productivity to the sheep and goats in the countryside. That had always been the primary aim of the Nimbkars. Superior bucks had been distributed to villages for mating to the local goats, in order to increase milk and meat output. For Deccani sheep – the main sheep breed in the state of Maharashtra – increased lamb production and improved resistance to worms in sheep were being introduced. This was done through providing rams that had the better characteristics bred into them through crossbreeding and selection. Bon and Chanda had made sure that the breeding plans at their research farm were relevant to the local community. They were pleased with the progress already made and they hoped to improve on it in future years. Gerald was proud to have played a part.

Both Bon and Jai, his wife, welcomed Gerald and Margaret to their home, where they were introduced to their warm, gracious family. They had another daughter, a medical doctor who worked in the hospital, and two granddaughters. A tour of the local town was organised for Gerald and Margaret. They were taken to the school for girls started by Bon and Jai. The children probably understood not a word of what Margaret was asked to say to them, but they had all been rehearsed to say 'goodbye' to her, and they shouted this in unison as the party left. The education of little girls was a passion of Jai's and she encouraged it wherever she went.

They also went sightseeing in the crowded market. Jai asked a local Muslim trader if all his children were at school. When he admitted, on questioning, that his girls were not, Jai told him in a strident tone that he must send the girls to school at once. It seemed that due to the fact that she and Bon were high-caste Indians, he being a Brahman, the highest caste, people accepted such lecturing without any argument.

The next day was spent at the home of Bon and Jai, where they enjoyed a meal served on the veranda. The house was dimly lit, the windows small, the furniture serviceable. The largest object in the room was a bookcase; they were a highly educated couple. Jai also did some writing and had had books published.

In the small yard at the back of the house were great piles of firewood. A man in white Indian clothing worked at a saw-bench. He did not look at them. The British couple stood still, looking up in surprise at the height and girth of the great mahogany trees at the back of the yard. Jai explained that these trees were her inheritance, and that they were worth a great deal of money.

A trip was planned by Chanda and Gavan. They wished to take their visitors to the Ajanta and the Ellora caves. These are one of the most important historic sites in India, found in the Western Ghats, inland from Mumbai and in the Deccan area. The trip was organised and the four of them set out on the long trip in Gavan's big jeep. They stopped at a village to buy fresh fruit, and arrived at the enormous house where they were to stay while they visited the caves. This magnificent building was the grace and favour house of the Chief Minister of Maharashtra. It had been offered to Chanda's father for the days that they were there. It was an enormous place, white and ecru coloured. Four grand arches faced the open landscape and a set of broad steps led down from the first floor. It was impressive, if not grandiose, and totally dwarfed the travellers. Two thin, white-clad men, permanent keepers of the place, were available with hot water and cooking facilities.

Next day the four of them arrived at Ajanta. From a distance they could make out the horseshoe shape of the barren hillside. The curving gorge dropping down steeply to the Waghora River. At one spot there was a waterfall cascading through seven pools. It was here in this secluded country that monks and others used to come to seek answers to the eternal questions of life and the causes of human suffering. The site was also on a trade route that linked the Arabian Sea to China.

The earlier caves had been carved out from the second century BC to the second century AD. The inside of the caves was stunning. Large rooms had been created as places of worship, both of Buddhism and Brahmanism. Here the stone was made to resemble the vaulted roofs of churches. In other caves the design was similar to that of Western monasteries, with little cells for the monks to sleep and pray surrounding a central rectangular space.

The second stage of cave carving occurred in the fifth century AD. The four were reduced to silence as they took in the incandescence of the colours of the paintings. There were pictures of men playing dice here, and a carved pink elephant in a lotus pond on the other wall. After this flowering of painting and carving, there was a decline in the ruling dynasty, and the caves were gradually abandoned, although still known by the locals.

Then, in April 1819, the caves were rediscovered by a soldier named John Smith, of the 28th Madras Cavalry. As he scanned the hillside for wild boar, he noticed the carved stone façade almost hidden by tangled shrubs. He and his companions scrambled down the hillside, and after all those centuries the caves were rediscovered. There was enormous interest and debate at the staggering feat of architecture. In later years these caves were declared a UNESCO World Heritage Site.

Chanda was proud and pleased to see the amazement on the faces of her friends. She went on to tell them that the minerals used for making the coloured paint had been obtained from local stone. The brilliant blue colour in the paintings was made from lapis lazuli, which had come all the way from Afghanistan.

Next, Chanda announced, they would visit the caves at Ellora, not too long a drive away. While the Ajanta caves were excavations of the hillside, those at Ellora, also now a UNESCO World Heritage Site, were carved out of the rock face of a hillside more than a hundred feet high. If anything they were even more astonishing than those at Ajanta.

The rock-cut architecture at Ellora had been started in the early seventh century AD and continued for more than two hundred years. There were twenty-nine caves completed under the Buddhist, Hindu and Jain faiths, which showed the tolerance and mutual respect of the different faiths of the time for each other.

The largest of the caves, known as the Kailasa temple, is the full hundred feet high, over three storeys. The courtyard contains high columns flanked by larger-than-life carved elephants. The holiest part of the temple rests on a high plinth. Sculptures abound and elaborate carvings decorate the walls and columns. Visitors are dwarfed by their surroundings. The art historian Percy Brown has described this excavation as 'the most stupendous single work of art ever executed in India'.

At this cave, they met up with a party of tourists. They heard a guide saying to them 'this richly carved temple is an example of the imagination, engineering skill, labour and perseverance of those who brought it into being. It is estimated that roughly three million cubic metres of rock had to be excavated to make Kailasa possible. The work proceeded downwards, the sculptors and carvers working on the top storey as the other masons continued to excavate the trenches.'

At one point of their tour, a party of modern-day monks in their orange robes approached. One of them was calling out to the few tourists viewing the caves to get out of the way. The swami, their leader, he shouted, did not want to look on women and must be kept apart from them. The monk who was speaking was bowed down as he walked, waving his right arm in front of himself whilst shouting 'Out of the way!', as the party of monks with their leader followed him in procession.

Chanda was enraged and told the monk that they were not moving and that the swami could look away or go somewhere else if he did not like women. This anger came straight from her staunch belief in the rights of women, something shared by all the Nimbkar family. Though Chanda was proudly Indian, her education, partly at Brown University in the USA, had made her intolerant of the male domination of Indian society. She was having none of this talk of women getting out of the way for a man.

They left the scene almost speechless. They passed quite a number of little wooden carts, each pulled by two bullocks and with enormous loads of straggly sugar cane. They stopped for a meal, very plain and bland, at a roadside café, and Chanda stopped at a water tap outside the place to clean her teeth with her fingers.

On the way home the crowds were thick. It was a Saturday, and a noisy wedding party was walking and laughing in the middle of the road through the village. They were accompanied by a thumping band of trumpeters and drummers. The charm and devil-may-care attitude of the people in their everyday lives delighted the visitors. They seemed to have qualities of natural dignity and unassuming confidence. Unlike back in Scotland, here in rural India there seemed to be no rat race, and no reason to hurry.

The sojourn in Maharashtra continued when Chanda's father and mother, Bon and Jai, invited Gerald and Margaret to accompany them on a trip to a hill station, Mahabaleshwar. In this land to the north of them, a few hours' drive would take them to the Club

Mahabaleshwar. This was a club founded during the British ruling of India for officers of the United Kingdom. Many of them were based in what was then called Bombay, and here they could find respite from the heat. Wives of officials of the British Empire were advised to go to the cool hills in the heat of the Indian summer.

Bon and Jai said that the air was cool there and the food was good. They added that there were lovely walks and beautiful birds to see. Bon quite proudly explained that he was the first Indian to be invited to be a member of this hill club, although now there were many Indians.

A few days later, they arrived at the hill station, where they found themselves on the edge of one of the few evergreen forests in India. The temperature there remained between seventy and eighty degrees Fahrenheit all year round, and during the British Raj of India it was an ideal place for a holiday away from the heat and stress of the city. The creation of a club was the idea of a group of officers of the British forces stationed in Bombay at the end of the nineteenth century. Today Mahabaleshwar is a popular holiday resort and honeymoon spot and an important pilgrimage site for Hindus.

For Margaret and Gerald this was a fairy-tale trip. It was January, and never had anywhere seemed in such contrast to Scotland in the winter. The red brick of the impressive buildings with their pointed arched frontage and the bright, well-tended gardens under the completely blue sky were a great invitation to relax. The two from Scotland were introduced to the manager of the complex, a charming, welcoming figure. This man, according to Bon, was a Parsee (a name for those from Persia in the past). The friendly staff soon showed them to separate bedrooms, part of a terrace in the grounds with steps down to the garden. Soon they found the hammocks under the old shady trees and were relaxing in the beauty of the place. Drinks were brought out to them by a smiling waiter. It felt as though they had arrived in paradise.

The next morning, their spartan, vegetarian Indian hosts called them for a seven-o'clock walk around the acres of garden and on to the golf course. Gerald was greatly amused by the golf 'greens' being called the 'browns', as they were sandy flat spaces without grass. The Nimbkars were keen birdwatchers, and on the party's return to the club for breakfast, they had to stop to identify the song of different birds, and to try to catch sight of each one.

One of the most interesting days of the visit took place when Bon announced that he was going to take them to the legendary source of the Krishna River in Old Mahabaleshwar, about one mile from the club. Not many people, he said, visited that place. They stopped at a wooded spot, where an ancient if unimposing ruin was located – the Temple of Mahadev.

Jai then led them down some steps to a terraced enclosure above a pool. At one end was a cast-iron statue of a cow with an open mouth from which water was slowly flowing into the pool below. It seemed a neglected looking thing, with the four of them the only visitors. But Bon wanted to share his excitement for this cherished statue. From the mouth of this animal, he told his two guests, sprang the water that is the source of the great Krishna River. He explained that, at first, the flow split into five great rivers, which joined up again eventually to form the great river. Jai must have noticed their incomprehension as she told them that this river is 1,300 kilometres long and flows through land where it means everything to the people of India. The delta region where it emerges at the Bay of Bengal is extremely fertile, she told them, for growing all kinds of crops.

Bon was keen to take his guests to a spot known as Ludwick Point. From there they had a magnificent view of the countryside for miles. An impressive monument stood there, with an inscription for all visitors to read:

This point now, by order of government, designated Ludwick Point in honor of his name. He reached here alone in 1824 after hours of toil through the dense forest: Here, therefore, as the most appropriate spot, this monument has, with the permission of government, been erected by his only son, R. W. Ludwick, H. M: Bombay Civil Service, Accountant-General of Madras in 1874.

There was no sign now of the dense forest but the tall stone monument brought history to life.

This had been an unforgettable break in their Indian sojourn. They had been shown, first by Chanda and Gavan and then by Bon and Jai, places of Indian history and culture off the normal tourist route, which not many visitors were privileged to see. Soon the time came when they had to say goodbye to Bon and Jai, to Chanda and Gavan, and to all the lovely family of Nimbkars. The thanks they extended to the Nimbkars were heartfelt.

The holiday continued for Gerald and Margaret in quite touristic fashion, taking in the usual sights such as the pink city of Jaipur with a ride uphill to the palace on the back of an enormous elephant. They also saw the stunning Taj Mahal, a marble mausoleum in Agra built by the emperor Shah Jahan in memory of his third wife.

From there they went on to Delhi and saw the Red Fort, an iconic symbol of India. With its massive red sandstone walls it is considered to be the zenith of Mughal architecture. It was the residence of the Mughal emperors for two hundred years. Next, they went on to visit the resting place of Ghandi, a moving experience in the wide garden-like place where they stood with thoughts of the great man.

Next day, their driver took the couple to a shop where their tailors would make an outfit for Margaret according to her measurements and have it ready next morning. While they stood in the large drapery shop, Margaret suddenly remembered that it was the twenty-fifth of January, Burns Day, a day of great national celebration in Scotland. She asked the men in the shop if they had heard of Robert Burns. They replied that they had not, and she explained that he was loved in Scotland and in many other countries. She went on to tell her guide that Burns lived more than two hundreds of years ago, and that he believed in the ideal that all men were equal and were brothers. Lines of one of his poems read, 'it's comin' yet for a' that, / that Man to Man the warld o'er / Shall brithers be for a' that'.

When she looked at the man who had become their friend on the long journey by car, she saw that there were tears in his eyes. She felt tears rising in her own eyes at the sentiment of the poem. Margaret said to him, 'But, of course, you have democracy here in India. You are the largest democracy in the world.'

He stood back a bit, saying to her surprise, 'Democracy! What use is that! India is still in a terrible mess! Corruption and problems of over-population and poverty everywhere!' Margaret turned away, having no answer to this outburst.

The journey home was fraught with trouble, mainly caused by Margaret feeling ill from food poisoning from a pasty she had eaten on a train and from the discomfort of a crowded airport at Delhi. A large party of American college students, Christian evangelists, who had been in South India and were on their way home, were taking the same plane to Frankfurt. They crowded out the area where they waited to board. An announcement came over the tannoy for people with young children and the aged come forward.

Margaret felt deathly as she sat hunched on her seat at the boarding point.

A friendly American student said kindly to her, 'You go ahead. You look aged!'

Margaret was both insulted and amused by this advice, but she gladly moved forwards and was one of the first with Gerald on to the huge plane.

A few hours were spent in Frankfurt, where Gerald had to wait while Margaret was put on a drip in the first-aid complex of the airport. Silently the medical people had listened to her story and led her to a hospital bed where she slept for several hours whilst being treated for dehydration. In the meantime Gerald waited anxiously on uncomfortable seating in one of the airport's vast halls until the time came for him and an awakened Margaret to take their onward flight to London and Edinburgh.

Chapter 35

Yak Conferences and New Visits

Gerald's Chinese colleagues and friends had clearly seen through the politics of the disastrous final meeting of the yak project in 1990 and placed no blame on Gerald for the break-up of that meeting. He had come to learn that they found value in his expertise and insights and were keen to have him back among them. Thus, much to his joy, as he greatly appreciated their friendship, he was reunited with his Chinese colleagues. Also, he had come to love the country – politics apart. Although these return visits spanned several years it seems best to relate them consecutively without intervening events, except one that is closely related to yak.

In 1993, Gerald was very pleased to receive an invitation to return to China for conferences and to lecture, following the end of the yak project. So this was indeed for him the start of happy times in a country whose people he had come to admire.

Gerald's return to China in 1994 was also helped by the good working relationship he had developed with Professor Cai Li during the yak project. He had described Cai Li on many occasions as a scholar and a gentleman – with a passion for yak. When the FAO suggested they might collaborate on producing a book on the yak, the idea was gladly taken up by both men. There were in existence many anecdotal and romanticised accounts of the yak but no comprehensive text in the English language on the yak's production characteristics, its products and its husbandry. The basis for such a book existed in a Chinese manuscript by Cai Li. It was translated into English by one of Cai Li's students, Xiangdong Zi, one of those who had benefited from the additional fellowships offered to the university in place of contentious extra laboratory equipment. To coincide with his particular interests, Zi had come to study in Edinburgh for a year and had attended a lecture course on animal breeding given by Gerald. They had also become friends.

But then, as Gerald euphemistically put it, 'the fun started'. He had been trained in Edinburgh in scientific methods based on thorough analysis of data. Stress had always been put on the necessity for careful interpretation of the results, and on their statistical significance. Cai Li, albeit an eminent scholar, belonged to an older generation of professors, not restricted to China, who had a looser view of what constituted evidence. The opinions of this older generation had been, in former times, accepted by the staff under them, usually without question. Their jobs depended on being on the right side of their bosses. For this reason, lengthy discussions ensued between Cai Li and Gerald on what to include and what to question. Cai Li, as it turned out, was fully amenable to the sometimes severe editing of his material and Gerald was also able to add much new information from published literature. The one part of Cai Li's manuscript that needed no change to its substance was his scholarly historical perspective on the yak. Cai Li claimed, without fear of contradiction, that the yak and the yak alone had made it possible for human civilisation to develop and exist on the high plateau and mountainous country of Western China. The yak had provided for every necessity for continued human existence: food, transport, fuel (from its dung), shelter and clothing.

In 1994 the first international congress on yak was held at Gansu Agricultural University in Lanzhou, one of the leading Chinese universities in its field. Gerald was invited to take part. This was also his first meeting with Cai Li since they had parted in Chengdu in 1990. They met as real friends, notwithstanding some of the critical correspondence during their collaboration on the yak book. Sadly, the two men were not to meet again, as Cai Li died in 1997.

Lanzhou was then a large but still expanding city surrounded by heavy industry, smoke belching from the chimneys of coal-fired furnaces. During the winter, Gerald was told, the pollution and smog were so great that it was difficult at times to see far, or to breathe comfortably. This was indeed one of the drawbacks of the rapidly growing economy and exploding sizes of the cities. Also, it was a consequence of the huge amounts of coal available to power the factories.

Gerald spoke to the congress of the opportunities that existed for the genetic improvement of yak and, importantly, some of the obstacles in the way to achieving this. He was also invited to chair some of the sessions, including one for animal breeding specialists. This is where he again came across the problem of the authority wielded by a department head in the university. Gerald had suggested that the mating of close relatives was an inevitable consequence of the way, traditionally at least, the dominant bull in the yak herd left more sons and daughters than any other bull in the herd. As a consequence his sons were likely to mate in subsequent years with some their half-sisters or even their mothers. And the dominant bull might even mate with some of his daughters. Just about everyone knows that inbreeding has harmful effects, Gerald contended. Little did Gerald anticipate that the department head would assert that inbreeding just did not occur with yak. That assertion ended any further discussion of that topic. A revelation came after that meeting, when two of the younger scientists approached Gerald to say that they had irrefutable evidence of inbreeding in yak. When asked by Gerald why they had not spoken up at the meeting, their answer was that their jobs depended on the goodwill of the professor.

This congress showed, as intended, that there was a wealth of research related to the yak, not only from China but from the several other countries with significant yak populations, notably Mongolia, Russia, Nepal, Bhutan and India. (China was home to around thirteen million yak and the other countries combined to another two million.) Much of that information had, in the past, been restricted largely to the countries from which it derived. This first international congress, also attended by a number of westerners, was a new way of spreading the message of contemporary research.

The congress was notable also for visits to a number of yak herds, which in Gansu Province presented an unusual sight as the predominant colour of the yak there is white, in contrast to the black or patchy hides that exist elsewhere.

Another visit warmed Gerald's heart and no doubt that of other animal geneticists. A new breed of yak was being developed at the Datong experimental station, but with a quite revolutionary idea. Wild yak, of which only a few thousand remained (after millions had been hunted to death for food and for trophies) were renowned for their larger size and vigour compared to the predominant domestic yak. The new breed was to be produced using the semen from a few captured wild yak bulls and by artificial insemination to produce crosses with domestic yak cows – and later to interbreed the crosses and select the best from among them for further breeding. The use of artificial insemination may seem commonplace nowadays as it is so widely used, but for the yak it needed a whole new technology to meet the reproductive characteristics specific to the animal. The end result was that this hybrid breed of yak (now called the Datong breed) had become a deservedly successful venture combining the best characteristics of two different types of yak.

The social side of the congress was fun – the Chinese know how to lay on a good feast and entertainment. Gerald also had one anecdote of some charm and poignancy. His interpreter at the congress was a bright post-graduate student, Ma Ying. Towards the end of his stay Gerald asked her if she would go shopping with him to Lanzhou so that he might buy some presents to take home. She agreed and it was done. Soon thereafter he was told by a colleague that she had previously booked a long bus journey to visit her parents in a province to the north. She saw her parents once or at most twice a year and had delayed her visit in order to please her guest. Gerald, and later Margaret, came to love Ma Ying almost as a daughter and they kept in touch with her as she rose in her profession – bright from the start and fortunately recognised as such.

The time Gerald spent in Nepal in 1996 was not a consultancy visit but resulted from an invitation to take part in a workshop, the Conservation and Management of Yak Genetic Diversity, to be held in Kathmandu. It was sponsored jointly by the International Centre for Integrated Mountain Development (ICIMOD) and the FAO programme for conservation and use of genetic resources in Asia.

It was the mountain development people from ICIMOD who were the hosts and organisers of the workshops at their headquarters in Kathmandu, Nepal. Altogether twenty-nine people were present from different countries involved with yak, though, surprisingly, Russian representation was absent. Gerald had met most of the participants before, but importantly among them were Han Jianlin and Ruijun Long from Gansu Agricultural University, who specialised in genetics and grassland science respectively. Both of these young Chinese scientists were already regarded as leaders in their fields and would become important collaborators for Gerald.

Many of the participants were asked to give papers and introduce discussion of various topics, all interrelated. There were country reports on the status of their yak populations and on products from them. Particular interest was shown in the development of yak cheese production in Nepal, which had been fostered with help and encouragement from Swiss experts. The cheese had found a niche market, especially among tourists, but was also marketed abroad. There was a suggestion that other countries with yak might consider a similar venture.

Gerald and Jianlin each presented a talk to a session devoted to yak genetic resources, to breeding strategies and to conservation of genetic diversity. Dan Miller, an American then attached to ICIMOD, and David Steane, of former Edinburgh days, on behalf of the FAO conservation programme, brought proceedings to a conclusion. From this intelligent and informative workshop should have followed action among the participating countries in terms of breeding, management and conservation policies, but whether this ever occurred was not obvious to Gerald. As so often happens, the initiatives of scientists and technocrats tend not to be what motivates the bureaucrats and the politicians.

During his visit to Kathmandu, Gerald once again took the opportunity to explore on foot. He delighted in the wealth of temples and shrines, the thrill of the crowded streets, the sight and smell of spices, and the distant snow-capped peaks of the Himalayan range disappearing in the evening sunset.

On the first day after his arrival, before he had met any of the other delegates, a chauffeur-driven car drew up alongside Gerald as he was wandering around the city and a figure called from the window of the car, 'Gerald, you are going the wrong way!' This was Dr Joshi, Director of the National Zoonoses and Food Hygiene Research Centre in Nepal. They had met at a previous conference. Gerald was sure he was not going the wrong way – though the map in his hand might have given that impression – but he happily joined his friend in the car and enjoyed an afternoon in one of the teahouses in the city. Margaret tells this tale when she wants to impress friends with how widely Gerald is known.

Gerald's return flight was via Calcutta in India. Arriving at six o'clock in the evening, when it was already beginning to get dark, he had seven hours before he was due back at the airport for his onward flight – a long time to wait at any airport and especially through much of the night. When he had booked his flights with the knowledgeable Trailfinders team they advised him to stay at Mrs Smith's guesthouse, Fairlawn, promising that it was an experience he would not forget. And so it turned out. A taxi took Gerald through the streets of Calcutta, where families were bedding down for the night on pavements and in the shelter of shop fronts, some cooking a meal on makeshift stoves. The taxi driver insisted that he remain at Fairlawn until 3 a.m. to return Gerald to the airport. He valued the fare and hoped for an extra large tip, which he confided would go to his son hoping to study at university.

Fairlawn was a villa built in 1783, though renovated since, surrounded by lush gardens and run by a widowed lady, Mrs Smith. She ran the place as though she were the matriarch of her family of guests. Her running of the guesthouse and her personal involvement in all its facets had been described as her 'obsession' – but a lovely obsession it was, as Gerald was to discover. Many famous figures from the literary and entertainment world had stayed at 'Mrs Smith's' including Tom Stoppard, Julie Christie, Clive Anderson and Ian Hislop, to mention just a few. And high on the list of former guests was a giant of a man described thus in an account of Fairlawn: 'A big (in every way) star who stayed here during the 1950s was the then-famous professional wrestler "King Kong". This giant of a man broke beds and chairs all over the hotel and consumed enough food for a small army.'

Gerald's short stay was less eventful. Mrs Smith welcomed him in person, showed him to an upstairs room which opened onto a balcony and gave instructions to come down to dinner when he heard the gong striking – a very old British custom and one he had not heard since his days in the Spooner household in Oxford, where this had also been a ritual. The meal was served to guests all sitting round a large table. There was no choice of food, but he recalls that he liked everything that was served. After the meal and some coffee and a little chat he repaired to his room to be woken from a short sleep just before he was due to return to the airport. Mrs Smith had attended to his every need and he was glad of this experience.

Qinghai province along with the Tibet Autonomous Region and part of Sichuan province forms a vast high plateau that rises from around 3,000 to 5,000 metres above sea level, interspersed by high mountains – the Himalayas to the south. This enormous area, the size of Western Europe, is home to the bulk of the domestic yak population in China as well as what remains of the wild yak.

The second international congress on yak was organised by the Qinghai Academy of Agricultural and Veterinary Sciences in Xining in 1997. It again drew specialists in yak production from a wide range of countries, including a few from Europe and the USA. As before, Gerald was invited to give talks at the congress.

The official language at these events was English, but for few of the delegates was that their first language. The majority were Chinese. Simultaneous translation was provided by students who, not surprisingly, sometimes struggled when a talk and the discussion that followed became too technical. Gerald took the precaution of preparing his talks verbatim and giving the interpreter a copy well in advance to allow for translation. What his talks might have lacked in spontaneity they made up for in accuracy in the translations.

Gerald's rehabilitation with his Chinese colleagues appeared to be complete as he was now able to say things, without being rebuked, that were sometimes critical of the way investigations with yak had been conducted. He was even able to question some traditional practice. For his sins, so he said, he was in danger of becoming regarded as a kind of *éminence grise*. Flattering though that was, he was determined not to become an old professor above argument. In any case, with a new generation of yak scientists starting to take over there was little danger of that.

When visiting China, Beijing was always the first port of call. Barely ten years since Gerald's first visit, the transformation in Beijing was mind-blowing. There were new buildings, roads and hotels, large shopping precincts with expensive imported goods, and taxis for the asking, something so absent before. Also, private cars were beginning to appear in large numbers, though not yet replacing the throng of bicycles. With the old parts of the city, the Forbidden City, palaces and old-time street markets still on the tourist trail, it was an exciting place. Ma Ying had married and was working for the Ministry of Agriculture. She and her husband invited Gerald out for a meal and chose to take him to McDonald's fast-food chain – a novelty and a treat for the folk of Beijing then. Surely a sign of the times.

In September 2000, Gerald was invited to the third international congress on yak, to be held in Lhasa. He was excited by the prospect of visiting Lhasa and seeing something of Tibet, a part of the Peoples' Republic of China where foreigners were not always welcome following China's tightened control over life there and the Dalai Lama's withdrawal to India. However, no restrictions were apparent, at least not to the delegates of the congress. The town was alive with people, many of them tourists. Lhasa lies 3,600 metres above sea level, but fortunately for Gerald he had not previously suffered from the thin air at these altitudes, provided he remembered to walk at a modest pace.

The first day or two were free for acclimatisation and therefore allowed sightseeing. The Potala Palace dominated the city, rising high above it with its thirteen storeys and seemingly endless flights of stairs. This was the former abode of the Dalai Lama, now a museum. It dates back to the seventh century but was rebuilt after a devastating fire in the seventeenth century. It was said to be the highest large building in the world, at 3,700 metres above sea level. Full of Buddhist artefacts and murals it is a stunning place to visit, with a breathtaking view of the snow-capped Himalayas in the far distance.

Gerald also visited the seventh-century Jokhang Temple. As he approached, he spotted men and women lying prostrate in the street in prayer. Another highlight was the walk through the old Barkhor district with its shops and street traders, the bustle brought into the twenty-first century by telephones and fax machines. The hotel where he stayed was on the other side of the Lhasa River from the main part of the town. The view of Lhasa from the bridge seemed to be a look back in time with women in colourful dress washing their clothes in the river and using it also as a social occasion.

The Lhasa congress was attended by seventy delegates from eighteen countries in addition to the hundred-strong contingent from China. As before, Gerald had been invited to present papers at the congress. New technologies based on understanding the molecular basis of inheritance were beginning to emerge. It was as yet unclear if these had much relevance to improving yak productivity, but the ideas suggested new opportunities. Gerald had sought the collaboration of Steve Bishop, a former colleague in ABRO, for this particular talk, as Steve had more intimate knowledge of the underlying techniques. Administrators from their offices, responsible for funding research, seemed keen to promote 'cutting edge' developments, such as in molecular genetics, without any clue how the new knowledge could benefit the yak industry. The talk tried to address these issues.

A second paper presented by Gerald caused perhaps more of a stir. Gerald had undertaken a survey (mostly by correspondence) of yak in 'non-typical' environments. To his surprise and delight he found that yak herds were performing successfully in places as far apart as North America and New Zealand, and at low elevations and in relatively warm climates at that. This went contrary to the received wisdom that yak were adapted, solely according to some, to high altitude and cold climates.

The advent of the yak congresses and the great amount of research and development presented at them made it clear that Gerald's original book on the yak, containing a preponderance of information gathered by Professor Cai Li, could now be superseded with vastly more up-to-date information. The FAO therefore encouraged a second edition. For this purpose Gerald invited the two eminent Chinese professors Han Jianlin and Ruijun Long to collaborate with him – they who had so impressed him at the meeting in Kathmandu. Both were of what Gerald called 'the younger generation', fully steeped in the ways of scientific investigation and the rigour needed for interpreting findings. Jianlin was a top-class geneticist, specialising in molecular genetics and widely knowledgeable about yak breeding. Ruijun was an expert on range management, environmental concerns and yak nutrition. The pair were just what was needed for a new, comprehensive book on the yak. Gerald himself contributed much new information from his search into the literature on the yak and, of course, from the many publications arising from the yak congresses. It was in the process of that search that he found the evidence that yak could do well in non-traditional environments, with possibly wider implications for the uses of the yak.

He coordinated the structure of the new book and provided its link to the earlier edition. The second edition of *The Yak*, under the authorship of Gerald Wiener, Han Jianlin and Long Ruijun, was published in 2003. It was more than twice the length of the first edition and differed from it markedly in both content and style. The debt owed to the late Professor Cai Li was clearly acknowledged.

Apart from the scientific sessions and the traditional Chinese feast – a meal of untold delicacies, many quite alien to the western palate – there was a trip to a yak breeding station in the majestic hinterland. At least as exciting to the foreigners was a visit to one of the many monasteries alive with red-robed monks, many quite young. It was difficult for Gerald to reconcile this with the supposed suppression of religious practice in China. There was no apparent evidence of that there.

Chapter 36

Events, Serious and Entertaining, in China

The trip in 2004 was special for Gerald as Margaret was able to accompany him on her first trip to China. It was to be a visit of work and sightseeing. The pair arrived in Beijing quite early and took a taxi to their downtown hotel. Gerald had stayed there before and followed the driver's route like a hawk with the help of a map. He wanted to avoid the three-mile detour made by a driver on a previous occasion and the ensuing argument over the fare. This time the driver was honest and deserved his tip. On the way to the hotel the taxi was held up in the crowded street of mid-town Beijing by several rickshaw taxis – tricycles with men peddling vigorously in front while the back two wheels held up a little cab. The one immediately in front of their taxi held a party of nine Chinese, jammed in close, their heads almost touching the roof of the cabin. This included a baby, sitting up in its mother's arms. Gerald told Margaret of how the streets were almost devoid of cars at the time of his first visit in 1988. Now they could hardly believe their eyes as they looked onto the four-lane highway, packed with cars, at times bumper to bumper.

The next day they met Ma Ying, the Chinese woman who had been Gerald's interpreter on his first conference visit to China. Margaret immediately found her as nice as Gerald had previously told her. With Ma Ying's husband, and their little son, Cong Cong, a child of about four, bright and mischievous, they visited the Forbidden City, awed by the traditional architecture and the vast spaces. The Hall of Supreme Harmony, with its double curved red roof fronted by three terraces of white marble, was indeed impressive as the centre of former imperial power. They were told that it was one of the largest surviving wooden structures in China.

The following day they were taken by their friends to the awe-inspiring Great Wall of China. From the spot where they stood they could see the wall stretching far into the distance. The place was crowded with sightseers, many of them Chinese parents with the one treasured child they were allowed to have. Margaret was a figure of curiosity to some of the children, as they did not often see plump, middle-aged Western women and wanted to be photographed alongside her. In the gift shop they admired the skill of delicate paper-cutting and bought two lovely pictures cut from red paper and mounted on an ochre background – one of a horse and one of a sheep.

For Gerald and Margaret the highlight of their short stay in Beijing was to be invited by Ma Ying and her husband to a meal at The Peking Duck, one of the most prestigious restaurants in Beijing. Accompanying them were Ma Ying's elderly parents. Alongside their table in the restaurant was a rack with a dozen or so dead ducks hanging by their feet, beaks pointing to the floor. Roasted duck made for a delicious meal in a friendly, happy atmosphere – except for the moment when Margaret, as the honoured guest, was presented with the prize piece of the feast – the beak of the duck. Though wary of causing offence, she declined this morsel, which was then eagerly taken by Ma Ying.

After these three fun-packed days in Beijing they flew to Chengdu. For Gerald this seemed like a homecoming – a return to the place of his consultancies but this time also a return to friends. The city itself had grown greatly as had so many other cities in China. However, it still retained much of its charm, not least the long line of stalls at the side of the wide avenue near the university campus. Here artists and craftsmen appeared in the early evening to display their skills and try to make a living.

The primary aim was of course to attend the fourth international yak congress. On this occasion Gerald was also called upon to lecture to staff and students interested in animal production at the Southwest University for Nationalities. Since Gerald's first visit, the university had grown explosively, with a large and beautifully landscaped campus and laboratories stuffed with the latest equipment. How the Mr Lin of yesteryear would have loved it and how all the aggravation over equipment in those earlier days might have been avoided had anyone only known what was to come.

Presenting a paper to the yak congress was no longer a novelty for Gerald. The choice of topic was more difficult as he had indeed become regarded, against his wishes, as an *éminence grise*. He chose, in his talk, to provide a perspective, as seen from afar, on past and present knowledge of the yak. Firstly he sought to discuss the basis for this knowledge, including cultural and social considerations, and secondly to consider environmental constraints and resources available in terms of yak breeds and breeding opportunities.

The talk was certainly well received but Gerald wondered if some of the older professors and those involved directly with yak production might have thought him an upstart. But no hint of that appeared, as far as he could tell. It was a well-attended meeting with representation from many countries, though once again there were no delegates from Russia or Mongolia either – two countries with not insignificant yak populations and research. One of the surprises for Gerald was the fact that every delegate was presented with a copy of the second edition of his book.

While Gerald was working hard at the conference, Margaret was assigned two female students to accompany her outside the university campus. The girls spoke quite good English and seemed to have been advised to adopt English names for themselves. One was Doris, the other Mabel, neither name any longer in fashion in the UK, but the names they had chosen sounded similar to their Chinese names. One of the trips that Margaret went on with the students was to the Chengdu Research Base of Giant Panda Breeding. Here she was to learn, as Gerald had on an earlier visit, of the serious study in the breeding of these beautiful animals. Gerald, on his visit years before, had been privileged to be taken to the panda 'maternity wing', not normally open to visitors, and shown a mother panda nursing her tiny offspring.

Another visit, this time organised by the congress secretariat, was to the famed Dujiangyan irrigation system, constructed in the third century BC. It was a scheme of unparalleled importance to the province, serving three purposes. First, it prevented damaging floods when the Min River swelled in spring due to melting snow from the mountains. It also kept part of the Min River open for the warships of the time, and by constructing secondary river flows it created a vast area of irrigated land that made Sichuan province into the food and fruit provider for all of China. For Gerald and Margaret it was awesome to see something so ancient and still of vital importance. The place had also been turned into a tourist attraction with an ornate entrance, great displays of flowers and of course the usual gift shop.

That evening, on their return, many delegates to the conference were gathered in the large entrance hall to have their blood pressure taken. This was to ensure they were fit enough to take the rarefied air of the high plateau to which they were to be transported. Luckily, both Gerald and Margaret were declared fit.

The next day after all the talking, eating and entertainment in Chengdu, came the drive into the high plateau – yak country. On the way to the yak project site they made a stop at the sublime Jiuzhaigou nature reserve, which Gerald had previously visited in 1988. Margaret also was delighted by the mirror lakes, waterfalls, amazing plant and bird life and majestic mountains. On this occasion there was more time to explore and wonder at this natural marvel. The place had now been developed into a fully functional tourist resort with hotels, restaurants, shops and small shuttle buses to take visitors to further-away parts of the park.

Two days at the park weren't long enough, but onwards they went in the convoy of several coaches to the yak territory and the yak farm. They were greeted by a display of herdsmen in traditional costume riding bareback on their horses, carrying flags and shouting greetings. Gerald and those accustomed to the altitude left the coaches to watch the display but Margaret and one or two others remained to view the herdsmen through the windows. They were supplied with some extra oxygen. The journey from low-lying Chengdu to the high plateau had been too fast for adjustment to the thinner air.

A large marquee had been erected for the occasion outside a newly built but as yet only half-finished hotel. That evening there was a party in the marquee, with plates of roasted yak for the guests. There was a team of TV interviewers who asked the visitors if they liked the taste. The party swelled in numbers as the evening wore on; people wandered around eating and chatting, both scientists and tribesmen. There was much drinking, but it was unclear what was being drunk. A traditional band started up, and the local men and women started to dance slowly, travelling around the tent in a circle. The dance consisted of walking and bobbing up and down, and many of the scientists joined in, though Gerald and Margaret abstained.

That night, senior members of the university on the trip became anxious about the health of Dr Wiener and his wife. They accompanied them to their bedroom, and stood and waited – with some others outside the bedroom door – for Gerald to undress for bed. They eventually realised, after being thanked for oxygen pillows and every other kindness, that the couple were not going to bed until they left the room.

The visitors were taken over the next couple of days to Nong Ri farm where the yak project had been sited. There was relatively little to see apart from the vastness of yak territory, the difficult terrain and the glorious mountains in the distance.

The visitors were also taken to a large, newly established commercial facility. With much public and private money poured into it, the intention was to breed superior yak hybrids for extra milk production and also collect yak milk from the local herders which was then be taken to a new processing plant. Yak milk and milk products, it was hoped, would find a niche market in large cities and abroad. A corresponding facility for processing yak meat was in the process of construction. Gerald and some of his Chinese colleagues were critical of the breeding plans in particular. They pointed to the obstacles to sustaining hybrid yak production on a large scale. They also wondered how cost effective a milk collection scheme, modelled on dairy cattle elsewhere, could be, given that surplus milk for sale would be available only during the short summer and early autumn season. The project was sensibly placed in the valley that had been pointed out to Gerald on his first visit as having a relatively dense yak population, atypical of other yak territories. Nonetheless, the yak herders would have to take relatively small quantities of milk quite long distances to the pick-up points on the only proper road in the valley. That too would add to costs. The criticisms were not well received by the managers of that project who had arranged the visit and the display. As far as Gerald could find out through correspondence in later years, the scheme did not live up to the expectations of the planners.

On the last day of the conference was the banquet – the great ritual at the end of each conference. About twenty large circular tables were set out with cutlery and glasses; dozens of bowls of Chinese delicacies were presented and the meal got underway. Large glasses of wine were provided and as soon as a glass was empty it was filled up. This was a noisy, laughter-filled occasion, with the Chinese hosts quickly letting their hair down.

Soon, the leaders of the Chinese team were striding round the tables of the visitors, noisily toasting each country's health. At the Wieners' table were two French scientists and also an American. They were slightly withdrawn from each other as there was some coolness between their countries due to disagreements over the invasion of Iraq by America. However, the two Brits toasted both 'La Belle France' and the United States with equal gusto, accompanied by their Chinese hosts.

As the evening moved on, someone got up and sang a song in Chinese. Then, as the younger delegates were noisily pleading with one of their table to sing, a young Chinese man got up on the platform and sang the Carpenters' hit song 'Yesterday Once More'. The young people, many of them students and interpreters, cheered and sang along. Gerald Margaret and the Americans at their table were gobsmacked by this performance in perfect English.

The visit to Chengdu was drawing to a close, and it became obvious that the university staff wished to treat Gerald and his wife before they left. A banquet meal in a local restaurant was arranged. This proved very enjoyable, if a bit drunken. When the Chinese drank, they sure knew how to knock it back. There was a competition between a top female student and her college lecturers to see who could down a half pint of beer the fastest. As she competed against several lecturers in turn, several pints were drunk. She won every time and suffered cheerfully, even proudly, the consequences of inebriation. This was a noisy and entertaining event.

Zi and Ma Li, who had previously undertaken fellowships for study abroad, arranged by Gerald, now professors at their university, wanted to make sure that Gerald and Margaret were entertained yet again on their last evening in Chengdu. Along with their wives, Zi's vivacious daughter and Ma Li's handsome young son, they had arranged for a lavish meal and a walkabout in a beautifully lit part of the town, with neon lights resembling parts of New York.

Returning to the university that night, they formed part of a merry group, slightly tipsy and with arms around each other, unusual for the Chinese, to be led by Zi in singing 'Auld Lang Syne'. This caused much amusement to the rest of the international group who were arranging their luggage ready for departure. One delegate, a research scientist from India, approached Gerald and asked if he would be kind enough to put his hand on his head and give him a blessing for the work he intended to undertake on yak reproduction on his return home. Gerald felt embarrassed at this new role bestowed on him, but courtesy demanded that he oblige. Some years later this research scientist wrote to Gerald to say that the blessing had allowed him to achieve a first in yak breeding – the production of a yak calf by embryo transfer.

Before leaving China, Gerald and Margaret had decided on a short holiday on their own. Their first stop was in Guilin, well known for the boat trip down the Li River, a day-long journey. On the boat, which held about forty tourists, were some seats on deck and a dining room below for lunch. The cruise was slow and the scenery stunning. The landscape on both sides of the river was marked by the famous limestone pinnacles that are the subject of countless Chinese paintings. It was a dreamy, sleepy and long journey, mesmerising in its beauty. Here and there were small flat-bottomed boats with ancient-looking fisherman using cormorants to catch fish for them – a timeless and enthralling scene. On the boat were people from many parts of the world, including, it turned out, the Mayor of Islamabad and his wife. Of course they invited Gerald and Margaret to visit them in Islamabad, but like so many such invitations it was probably just a formality and never taken up.

At the end of the trip, the boat docked at a riverside village, obviously a tourist spot. There was a street lined with little shops and booths selling souvenirs, the same as at ports the world over. Surprisingly there was a stall selling T-shirts with a picture of the bearded bin Laden on the front. This was not long after the atrocity of the Twin Towers' destruction in New York and they thought this to be in bad taste.

Later the couple visited Shanghai, a city grown out of all proportion in size and modernity from the city in which Gerald's family in the USA had found refuge from persecution in Nazi Germany. As it was a public holiday, the crowds were incredible. Skill and concentration were required to get through the people in the street, and not to lose your partner or group in the melee.

The last visit of this brief holiday was to Suzhou, not far from Shanghai. It is a large city beside a lake with magical canals, bridges, gardens and a stunning old pagoda on top of a hill.

Gerald and Margaret flew back to Beijing and had a couple more days of sightseeing. Thanks to Ma Ying and her husband they were taken to the Summer Palace outside Beijing, first built more than two centuries earlier and boasting a magnificent imperial garden. It was a fitting end to a memorable trip; for Gerald it was a reminder of how much he admired China, for Margaret an introduction to a world she had formerly only imagined.

Chapter 37

Later Years

At home in Biggar, life might have seemed dull after all that travel. In fact there could have been even more travel if Gerald had not scrupulously turned down a number of offers because he thought someone else was better qualified for the specifics of the consultancy on offer.

In addition to his continuing visits to Roslin to complete analysis of his accumulated research data and its publication, his consultancy work and his lecturing commitments, Gerald had started a new pursuit around the time of his statutory retirement from ABRO in 1986. He was invited to take over, from Ian Mason, the course of post-graduate lectures on animal breeding at the Centre for Tropical Veterinary Medicine at the University of Edinburgh. He greatly enjoyed this role and continued in it for several years. He was very pleased by the keenness of the students, many from overseas countries and often supported financially, in those days, by the British Council. This course of lectures then formed the basis for his book *Animal Breeding*, published by Macmillan in their Tropical Agriculturist series. The book met with considerable success, having to be reprinted after its original print run. It was also noticed by Chinese colleagues and translated into Chinese as, it was said, no 'modern' textbook on animal breeding and genetics existed in China at that time. It was used for students in China for a number of years. The book was also, subsequently, translated into French.

In these later years Gerald continued to correspond with former colleagues, particularly those in China and most often with Han Jianlin. He also continued to receive scientific papers from journal editors for review. But increasingly he felt insufficiently up to date in some of the more technical aspects of research and started to refuse the job of referee, even though from time to time it brought back happy memories of the past.

Gerald and Margaret paid several more visits to the USA to visit his family there. Aunt Thea, though visibly ageing, was still full of fun and continued to be particular about her appearance – she had her nails manicured every week. Gerald also got to know his two brothers, Jerry and Pete Wiener, better, and the new generation of nephews and nieces.

During one trip in December they were entertained at Pete and Judie's house. To the visitors from the UK it was an odd experience that, amidst Christmas decorations and presents strewn around, Kurt, the husband of Gerald's stepmother, Ursel, the only strict Jew among this religiously diverse crowd, was reciting Hebrew prayers for the festival of Hanukkah.

In Biggar, Gerald continued his involvement with community affairs and, to his own amazement, let Margaret persuade him to take lessons in the game of bridge and then join the local and very active bridge club. He had forsworn the game some sixty years earlier but now he started to enjoy it. Most of the other players in the bridge club had played the game for many years, some most of their lives, so winning was rare – but he enjoyed the challenge. Margaret, being of a more sensitive nature, disliked losing.

When Gerald reached the age of seventy-five he decided it was time to retire from active duties as an elder of the Kirk. There were too many old people in the church relative to the younger generation – but in a rural community like Biggar this was not quite as big a problem as for some city congregations.

In 2006 he was surprised by the arrival of a high-powered delegation from the Southwest University for Nationalities in Chengdu, the university he had been attached to for the yak project and which had made him one of their honorary professors. The seven members of the party included the principal of the university and Xiangdong Zi, now a professor, who had studied for a year in Edinburgh financed by one of the FAO fellowships Gerald had been allowed to allocate to students back in 1988–89. Gerald and he had, of course, met again in Chengdu some years earlier, but it was good to see him now in this senior role. The group were on a mission to attract collaboration from a number of universities in England but had detoured to Edinburgh in their hired minibus specifically to meet up with Gerald. He felt duly honoured by this gesture.

Gerald and Margaret met the group at a university residence in Edinburgh prior to taking them out for a meal at the Sheraton Hotel. They were surprised when the group formed a semi-circle round them and the university principal made a short speech of appreciation for this get-together and proceeded with their charming custom of placing silk scarves around Gerald's and Margaret's necks. The meal and the rest of the evening turned out to be thoroughly convivial and proof to Gerald that his visits to China had been worthwhile.

That year Gerald also celebrated his eightieth birthday. Andrew, his son, arranged a party for Gerald at the luxurious Grove Hotel in Hertfordshire to suit those friends and relations living in that part of England. It was a pleasure that Gerald's childhood friend Hardy Seidel and his wife, Ruth, were able to travel from London to be there.

The following year there was a visit from another delegation, perhaps less prestigious than the previous one, but no less welcome. Ma Ying, the young woman Gerald had first met at Gansu Agricultural University when she was a post-graduate student and who had been so welcoming to Gerald and Margaret in Beijing, arrived in Biggar with a party from China. Ma Ying, now their firm friend, had risen to a position of responsibility in the Ministry of Agriculture in Beijing.

Ma Ying was leading a party of officials in agriculture from China to the UK to establish cooperation with government bodies in the UK. She thought it would be possible for them to detour to Biggar to meet with Gerald and Margaret. And so it happened that the party of five came for much of a day to their home in late August. Margaret made a meal for them and they enjoyed a rare opportunity to be in a private house and not to see everything just from hotel rooms and offices. One of their party commented that in China a house of that size would be occupied by more than two people. Their hosts agreed that the house was too large for them – but as the former schoolhouse for the headmaster of the high school it had been built for a larger family and for entertaining.

The party enjoyed their brief visit and some of them were particularly impressed by what seemed to them a garden large enough to grow most of the food for the household but, as they saw, given over mostly to shrubs, flowers and grass with only a small area for growing vegetables. An apple tree was, by good luck, laden with fruit and they took delight in picking some to eat. This was another happy event for Gerald to add to his China experiences.

The Wieners moved to Inverness in the summer of 2008. They wanted to be near the children settled in that area of the Scottish Highlands. As they grew older, they felt it would be good to have family within easy visiting distance. It proved a good move and the couple, after long and busy lives, were now starting to relax.

One event occurred soon thereafter that took Gerald right back to his boyhood in Berlin and his early years in the UK. Gerald's lifelong friend Hardy Seidel was to celebrate his diamond wedding with his lovely wife, Ruth. Gerald and Margaret happily accepted the invitation to the event in London.

Memories started to flood back. Hardy, fourteen months older than Gerald, had but a brief period at school in pre-war London. He was then thrust into helping Rudi, his father, in his business, manufacturing gifts. It is another story of a refugee entrepreneur. With a business partner, Rudi was turning basic, unadorned objects like powder compacts into things of beauty that they hoped every woman would want. That business did not survive the war years but it made Rudi think that companies and firms, both large and small, might be attracted to the idea of giving small gifts to their customers as tokens of appreciation. And so Anglo Fancy Products was born. It turned into a successful enterprise long before large competitors arose to turn this idea into big business, so that nowadays the 'give-away item' is commonplace.

Hardy used to travel up and down the country to drum up business with a van loaded with sample goods from Anglo Fancy Products. But he never ventured much north of Manchester where he seemed to imagine the country stopped, and he never visited his friend in Scotland.

Hardy had been madly keen on Ruth from the time she was only sixteen years old. He said that she played hard to get. Gerald was a frequent visitor to the Seidels and having met Ruth as a young woman he agreed with Hardy that Ruth was beautiful. A grand reception was held for their diamond wedding at the home of their older daughter, Jacqueline. Gerald was asked by Ruth to say a few words as no others could claim a friendship with Hardy back to their boyhood days in Berlin. Gerald recalled not only those days but the many happy visits he had made to their home in London. Oft times, Gerald would be 'lectured' by his slightly older friend on dos and don'ts with girls and with money. Hardy was much more astute than Gerald in the world of finance. Gerald also recalled joyful times spent with Hardy's parents. Herta, Hardy's mother, had been like a second mother to Gerald – without the admonitions he got from Luise. How Hardy's parents would have loved this occasion with their family, but of course they had passed away years earlier leaving Hardy, with the help of daughters Jacqueline and Karen, to further expand the business that his father had started.

Hardy always teased Gerald with how hard he had to work to make a living whilst Gerald was luxuriating at university and then in a job that allowed him to 'see the world' – a cushy job, as Hardy put it.

Sadly Hardy died in 2010. A couple of years later, Ruth, accompanied by her daughter Karen, paid her first ever visit to Scotland. Gerald and Margaret were delighted to see them. Their visitors had to admit that the Highlands were not a 'frozen waste' – a thought that had kept Ruth and Hardy away for all those years.

In 2009, Gerald had contacted the farm manager of a large landowner in Scotland, the Duke of Buccleuch, to ask if he and his employer, the duke, would give some thought to importing yak with idea of the crossbreeding them with Highland cattle.

It was known by Gerald, from his research, that there were many successful yak herds in other countries, including several in Europe, with varying climates and environment. Several parts of Scotland, he felt, would be ideal for yak and a niche market for yak products could result.

A meeting took place in Inverness between the duke and Gerald. However no decision was taken on that day. The duke feared that his farm manager would disapprove of something so contrary to tradition.

Han Jianlin visited and stayed with Gerald and Margaret in Inverness twice, the first time on his own and the second accompanied by his wife. It was on these occasions that they discovered he was an expert not only on the yak but also a knowledgeable connoisseur of whisky. Ruijun Long, one of the other collaborators on *The Yak*, took time out from a conference in Perth, some

110 miles from Inverness, to visit Gerald and Margaret for part of a day. Gerald enjoyed these links to his former work. He also continues to have frequent correspondence with the Nimbkars in India. Chanda keeps him abreast of her work as a now widely respected animal geneticist and breeder, and Bon, though now, like Gerald, getting old, continues to have ever expanding ideas for his research institute.

Gerald has kept in touch with his family in the USA. Sadly Thea and Ursel, as well as Judie Reid, have passed away, but others are still going strong, though Hans and Vera, like Gerald, complain of the side effects of getting old. Also, a new, even younger generation of the family has been born that Gerald has not met. To Gerald and Margaret's delight his half-brother, Jerry, and his wife, Mari, pay regular visits to them in Scotland, roughly every two years. Jerry is also Gerald's most regular correspondent. Judie Wiener, wife of Gerald's other half-brother, Pete, came on one visit. So the bond with the family in the USA continues strongly.

Two things occurred in 2015 that took Gerald back to his earliest times in the Animal Breeding Research Organisation. The first of these was a visit to him in Inverness by Clare Button, archivist to a project funded by the Wellcome Trust to catalogue and conserve the collections and information that tell the story of animal genetics and animal breeding at the University of Edinburgh and ABRO, back to its earliest beginnings in the mid-nineteenth century. Clare, complete with dictation machine, came to interview Gerald to add his recollections to the many others. The title for this massive project is, appropriately, 'Towards Dolly'.

The second event held greater poignancy. Gerald and Margaret, whilst on a Baltic cruise holiday had the opportunity to visit, in Helsinki, his erstwhile colleague and friend of many years Kalle Maijala, the man who had introduced Finnish Landrace sheep to him back in 1961 with the far-reaching consequences described in an earlier chapter. Sadly, Kalle was suffering from advanced

Parkinson's disease, confined to a wheelchair in a nursing home and was barely able to speak. Kalle's wife, Kirsti, took Gerald and Margaret to see him in his room at the home. Signs of the latest of many honours bestowed by Finland's academia on their highly esteemed professor emeritus were all around. Kalle had been briefed about Gerald's visit and, according to Kirsti, he not only recognised his friend from Scotland but understood the conversation and reminiscence. He also enjoyed, as did Gerald and Margaret, the lavish picnic that Kirsti had prepared. Gerald could not altogether contain the strong emotions evoked in him on seeing his friend, the formerly enthusiastic and vigorous young scientist, so severely disabled. As they parted, Kalle held Gerald's hand very firmly. Gerald felt this as a wonderful moment. Sadly, Kalle passed away on 4 March 2016.

Gerald turned ninety in April 2016. He continues to go to the local Church of Scotland about once a month. He also goes to bridge lessons and enjoys the occasional game. Short holidays are now enjoyed nearer home, as travel has become more tiring and difficult to endure as the years pile on.

Like most people he hopes he has succeeded in life and made, possibly, some difference for good. To end there is one more flashback to Gerald's life as scientist and animal geneticist. It makes a fitting postscript.

After an interval of ten years the Chinese government finally agreed to the fifth international congress on yak, to be held in Lanzhou, Gansu province, in August 2014. This time the congress was hosted under the auspices of the Lanzhou Institute of Husbandry and Pharmaceutical Sciences and of the Lanzhou University – not of the Gansu Agricultural University of the first congress.

No official reason was given why this permission was so long delayed. Rumour had it that the government was reluctant to allow Tibetan scientists to mingle freely with those from other parts of China and from foreign countries. This had to do, so it was thought, with international criticism of the manner in which China was administering its Tibetan region.

Gerald was unable, for health reasons, to accept the invitation to attend this congress. All his instincts had been to go and meet again many friends and colleagues and to take part in the discussion on yak that he had come to enjoy so much. The talk he had wanted to give, a kind of reminiscence, was presented in part by his friend Professor Han Jianlin, alongside Jianlin's own paper. When the proceedings of the congress were published, the opening article to appear was the full text of the talk he had prepared with the title 'Looking back and looking forward – as seen from the outside' by Gerald Wiener.

To be honoured in this way was not just a complete surprise to Gerald but made him feel both privileged and emotional. Surely his time in China had not been wasted. Through the yak books and talks and lectures outside China he had also helped to bring to wider audiences the extraordinary attributes of the yak and its potential for use in harsh conditions in different parts of the world.

Index